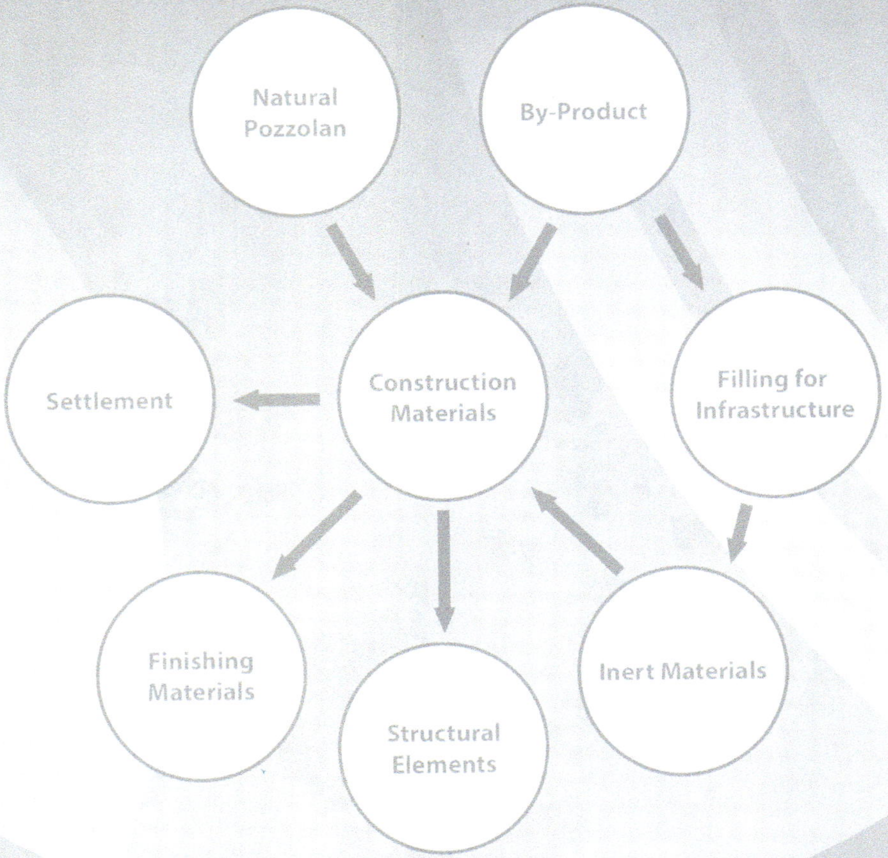

Handbook of Innovative Adhesive Technology

Handbook of

Innovative Adhesive Technology

Biodegradable and Cost-Effective Binder for Built Environment and Engineering Sciences

edited by

Mehmet Serkan Kirgiz

JENNY STANFORD
PUBLISHING

Published by

Jenny Stanford Publishing Pte. Ltd.
101 Thomson Road
#06-01, United Square
Singapore 307591

Email: editorial@jennystanford.com
Web: www.jennystanford.com

British Library Cataloguing-in-Publication Data
A catalogue record for this book is available from the British Library.

Handbook of Innovative Adhesive Technology: Biodegradable and Cost-Effective Binder for Built Environment and Engineering Sciences

ISBN 978-981-5129-51-9 (Hardcover)
ISBN 978-1-003-59892-3 (eBook)

Contents

Preface xv

1. Starch-Based Compound for Eco-friendly and Sustainable Construction Adhesive and Binder System **1**

Mehmet Serkan Kirgiz

1.1 Introduction 2
1.2 Approaches of Mixing of Starch-Based Adhesive Systems for Eco-friendly Construction Materials 3
1.3 Structure and Features of Starch 6
1.3.1 Amylose and Amylopectin of Starch 8
1.3.1.1 Amylose 8
1.3.1.2 Amylopectin 8
1.4 Other Compounds in Starch Powder 9
1.4.1 Different Types of Starch and Their Ratio of Amylose to Amylopectin 9
1.4.2 Eco-friendly Construction Materials with Starch 10
1.4.3 Unit Mass 10
1.4.4 Thermal Conductivity 14
1.4.5 Compressive Strength 16
1.4.6 SEM Analysis 19

2. Self-Compacting Composite Containing Mine Tailing and Fortified by Nanocarbon Dot for Manufacturing of Cost-Effective and CO_2-Neutral Material in Construction and Infrastructure **27**

Mehmet Serkan Kirgiz, André G. D. S. Galdino, Ayub Elahi, and Atizaz Ali

2.1 Introduction 28
2.2 Materials and Methods 29
2.2.1 Materials 29
2.2.2 Pretreatment to Improve the Effectiveness of SCM and Nanocarbon 29
2.3 Results and Discussion 30

2.3.1 Chemical Component 30
2.3.2 Compressive Strength 31
2.3.3 Splitting Tensile Strength 32
2.3.4 Flexural Strength 33
2.4 Conclusion 35

3. Valorization of Nanocarbon Dot–Based Green Binder Systems in Terms of Earthquake Engineering, Reinforced Concrete Building, Applied Mechanics, Transportation Engineering, and Their Applications 39

Mehmet Serkan Kirgiz, André Gustavo de Sousa Galdino, Raul D. S. G. Campilho, Rajesh Kumar, Konstantinos G. Kolovos, and Vinod M. Kumar

3.1 Introduction 40
3.1.1 Aim of This Study 41
3.2 Experimental Procedure 41
3.2.1 Preparation of Green Cement and Green Composite Materials for Testing 41
3.2.2 Determination of Water Absorption 41
3.2.3 Determination of Compressive Strength 43
3.2.4 Determination of Flexure Moment Strength 44
3.3 Results 44
3.4 Discussions 44
3.4.1 Water Absorption of Material 44
3.4.2 Changings of Compressive Strength Gain of Material 45
3.4.3 Changings of Flexural Moment Strength Gain of Material 46
3.5 Conclusion 47

4. Some Technological Properties of Hydraulic Construction Lime and Grouting Incorporated with Chopped Carbon Nanotube Fibers for Manufacturing Cost-Effective and CO_2-Neutral Materials in Cleaner Built Environment 51

Mehmet Serkan Kirgiz, Andre G. D. S. Galdino, John Kinuthia, Raul D. S. G. Campilho, Roberto A. Gonzalez-Lezcano, and Konstantinos G. Kolovos

4.1 Introduction 52

4.2 Materials and Methods 53
4.2.1 Materials in Making Grout Sample 53
4.2.2 Preparation of Grouting 53
4.2.3 Testing 54
4.3 Results and Discussion 54
4.3.1 Chemical Components of CL90S 54
4.3.2 Physical Properties of CL90S 56
4.3.3 Apparent Porosity 57
4.3.4 Water Hydrophobicity 59
4.4 Conclusion 60

5. Transforming Waste Oil Sand Ash into Construction Material Using Net-Zero Energy **65**

M. Hesham El Naggar, Mehmet Serkan Kirgiz, André Gustavo de Sousa Galdino, Roberto A. Gonzalez-Lezcano, Raul D. S. G. Campilho, and Konstantinos G. Kolovos

5.1 Introduction 66
5.2 Application of Treated OSW in Controlled Low Strength Materials 69
5.3 Materials and Methods 69
5.4 Results and Discussion 70
5.4.1 XRD, SEM, and EDS of TOSW 70
5.4.2 Particle Size Distribution of TOSW 71
5.5 Conclusion 72

6. Ecological, Sustainable, Green Brick Materials: Unfired Manufacturing System, Some Features, and Usage for Greener Built Environment and Construction Technology **73**

John Kinuthia, Mehmet Serkan Kirgiz, André Gustavo de Sousa Galdino, Raul D. S. G. Campilho, Konstantinos G. Kolovos

6.1 Introduction 74
6.2 Durability 83
6.3 Industrial Additives 84
6.3.1 Ground Granulated Blast Furnace Slag 85
6.3.2 Pulverized Fuel Ash 86
6.3.3 Waste-Paper Sludge Ash 87
6.3.4 Agricultural Additives 87

6.3.5 Construction Using Clay-Based Systems 87
6.3.5.1 Storage and construction 88
6.4 Concluding Remarks and Future Trends 88

7. Adhesively Bonded *T*-Joint Optimization by Cohesive Models: Effect of Adhered Material, Adhesive Type, Geometry, and Bonding Technique 95

J. P. M. Lopes, Raul D. S. G. Campilho, and Mehmet Serkan Kirgiz

7.1 Introduction 96
7.2 Materials and Methods 99
7.2.1 Description of *T*-Joint Models 99
7.2.2 Adherends and Adhesives 100
7.2.3 Model Pre-processing 101
7.3 Results 102
7.3.1 CZM Validation 102
7.3.2 Adhesive Type 104
7.3.3 Geometrical Parameters 105
7.3.4 Dual Adhesives 107
7.3.5 Composite Adherends 109
7.4 Conclusion 110

8. High-Performance Concrete Made of Seawater and Cured in Seawater to Evaluate Hydration Composition and Porosity with Analysis of SEM, Petrography, and Mercury Porosimetry in the Lack of Fresh Water of Cleaner Built Environment 115

Erniati Bachtiar, Fatmawaty Rachim, Ritnawati Makbul, Arbain Tata, Muhammad Syarif, Mehmet Serkan Kirgiz, André Gustavo de Sousa Galdino, Konstantinos G. Kolovos, and Andi Yusri

8.1 Introduction 116
8.2 Materials and Method 118
8.2.1 Materials 118
8.2.2 Research Methods 119
8.3 Results and Discussions 121
8.3.1 Hydration Products and Changing in Friedel's Salt ($3CaO{\cdot}Al_2O_3{\cdot}CaCl_2{\cdot}10H_2O$) and Chlorine ($Cl_2$) with XRD 121

8.3.2 Microstructure and Morphology of HSC with SEM 124
8.3.3 Porosity of Concrete 126
8.3.4 Absorption of Concrete 128
8.4 Conclusion 129

9. Portland Cement and Starch-Based Adhesives: A Comprehensive Review **133**

Konstantinos G. Kolovos and Mehmet Serkan Kirgiz

9.1 Introduction 134
9.2 Starch Chemistry and Properties 134
9.3 Starch-Based Adhesives 135
9.3.1 Starch and Cement-Based Materials 136
9.3.2 Starch in Concrete 138
9.3.3 Starch as Water-Reducing Agent in Concrete 140
9.4 Conclusion 141

10. Properties of Innovative and Eco-friendly Pavement and Structural Materials Using Waste Marble Sludge Dust, Rice Husk Ash, Coir Fiber, Starch-Based Binder, and Carbon Quantum Dots **147**

Mehmet Serkan Kirgiz, Jamal Khatib, Muhammed Syarif, Erniati Bachtiar, Olatokunbo M. Ofuyatan, Touhami Tahenni, Krishnaraj Ramaswamy, Ahmed Ashteyat, Ayub Elahi, Khairunisa Muthusamy, and Gideon O. Bamigboye

10.1 Introduction 149
10.2 Applications of Pavement and Structure Materials Containing S-BDB, RHA, WMSD, CF, and CQDs 150
10.3 Chemical Properties of S-BDB, RHA, WMSD, CF, and CQDs 151
10.4 Physical Properties of S-BDB, RHA, WMSD, CF, and CQDs 151
10.5 Preparation and Testing of Innovative Composite and Control Composite 153

10.6 Effect of RHA, WMSD, CF, and CQDs on the Fresh Properties of Pavement and Structural Materials Made of S-BDB 153
10.6.1 Setting Time 154
10.6.2 Rheology with Slump 156
10.7 Effect of RHA, WMSD, CF, and CQDs on the Hardened Properties of Pavement and Structure Materials Made of S-BDB 157
10.7.1 Density 157
10.7.2 Compressive Strength 158
10.7.3 Splitting Tensile Strength 160
10.8 Conclusion 162

11. Behavior of Bacterial Strain over Cementless Geopolymer Concrete for Sustainable Development 169

A. Chithambar Ganesh, M. Vinod Kumar, A. Navya, G. Dasaradha Rami Reddy, S. R. Ragavi, C. Navin Gowtham, and Mehmet Serkan Kirgiz

11.1 Introduction 170
11.2 Materials 171
11.3 Culturing of Bacteria 172
11.4 Mix Design and Casting 172
11.5 Results and Discussion 173
11.5.1 Compressive Strength 173
11.5.2 Splitting Tensile Strength 174
11.5.3 Flexural Strength 175
11.5.4 Water Absorption 175
11.5.5 Self-Healing Property 177
11.6 Conclusion 178

12. Geopolymer Bricks Using Industrial Wastes to Reduce CO_2 Footprint 183

A. Chithambar Ganesh, R. Mohana, M. Vinod Kumar, V. Prem Kumar, and Mehmet Serkan Kirgiz

12.1 Introduction 184
12.2 Materials and Methodology 185
12.3 Results and Discussions 186
12.3.1 Effect of GGBS–Iron Slag Content 186
12.3.2 Effect of Na_2SiO_3 and NaOH 189

12.3.3 Effect of Molarity 191
12.4 Conclusion 193

13. Natural Rubber Latex–Substituted Bitumen Binder for Road Construction: A Review **197**

Trinity A. Tagbor, Mehmet Serkan Kirgiz, Theresah Osei, Latifatu Mohammed, Ammishaddai Ofori-Nyarko, Festus Mensah, and Samuel Obiri

13.1 Introduction 198
13.2 Materials and Methodology 199
13.3 Natural Rubber 200
13.3.1 Use of Natural Rubber 200
13.3.2 Physical Properties of Natural Rubber 201
13.3.3 Chemical Properties of Natural Rubber 201
13.3.4 Natural Rubber as a Binder 202
13.4 Bitumen 202
13.5 Natural Rubber as a Binder in Bituminous Mixes 203
13.6 Challenges 207
13.7 Conclusion and Recommendations 208

14. Agro-Industrial Supplementary Cementitious Materials for Realizing Sustainable Development Goals of the UN and Net-Zero CO_2 Emission **215**

Ahsan Waqar, Naraindas Bheel, Mehmet Serkan Kirgiz, Muhammad Basit Khan, and Muhammad Shoaib Mansoor

14.1 Introduction 216
14.2 Agricultural Wastes 218
14.2.1 Millet Husk Ash 218
14.2.2 Wheat Straw Ash 219
14.2.3 Corncob Ash 221
14.2.4 Palm Fuel Oil Ash 223
14.2.5 Groundnut Shell Ash 225
14.3 Industrial Wastes 227
14.3.1 Ground Granulated Blast Furnace Slag 227
14.3.2 Limestone Powder 229
14.3.3 Silica Fume 231
14.3.4 Sewage Sludge Ash 233
14.3.5 Quarry Dust 235

15. Natural Rubber Latex and Its Effects on Some Properties of Bitumen for Cleaner Built Environment **245**
Theresah Osei, Trinity A. Tagbor, Johannes A. M. Awudza, and Mehmet Serkan Kirgiz
15.1 Introduction 246
15.1.1 Objective 248
15.2 Materials and Method 248
15.2.1 Bitumen 248
15.2.2 Natural Rubber Latex 248
15.2.3 Sample Preparation 248
15.2.4 Laboratory Analysis 249
15.2.5 Aging Index 250
15.2.5.1 Penetration index and temperature susceptibility 250
15.2.5.2 Specifications by Ghana Highway Authority 250
15.3 Results of Analysis 251
15.3.1 Penetration Point 251
15.3.2 Softening Point Temperature (°C) 251
15.3.3 Viscosity Analysis 252
15.3.4 Specific Gravity Test 253
15.3.5 Temperature Susceptibility and Penetration Index 253
15.3.6 Flash Point, Aging, and Viscosity of Selected Blends 253
15.3.7 FTIR Analysis 254
15.4 Discussion 254
15.4.1 Penetration Point 254
15.4.2 Softening Point Temperature 255
15.4.3 Viscosity 256
15.4.4 Specific Gravity 256
15.4.5 Temperature Susceptibility and Penetration Index 256
15.4.6 Flash Point, Aging, and Viscosity of Selected Blends 257
15.4.7 FTIR Analysis 257
15.5 Conclusion 258

16. Analysis of E-Waste Containing Carbon and Other Components as Composite Material for a Cleaner Built Environment and Realizing UN Sustainable Development Goals **261**

M. Vishnu Priyan, R. Annadurai, Mehmet Serkan Kirgiz, André Gustavo de Sousa Galdino, Raul D. S. G. Campilho, Roberto Alonso González-Lezcano, and Konstantinos G. Kolovos

16.1 Introduction 262

16.1.1 Sources of E-waste and Its Stance in the World 263

16.2 Materials and Methods 263

16.3 Results and Discussion 265

16.3.1 Specific Gravity 265

16.3.2 Water Absorption 265

16.3.3 Compressive Strength 267

16.3.4 Split Tensile Strength 269

16.3.5 Flexural Strength 271

16.4 Conclusion 272

Index 279

Preface

This book represents the significant effort that has been put in and the environmentally conscious approach that has been used to synthesize the vast knowledge gained on the use of alternative materials in construction.

It would be fair to say that there has been a notable increase in global cement production over the past decade. This trend represents a significant factor affecting technological development and the updating of manufacturing facilities in the cement industry. It can be said that the technology currently used to produce cement clinker has some negative environmental consequences as it is known to consume a great deal of energy and natural resources as well as emit pollutants. This book presents a new approach to the production of a biodegradable and cost-effective binder that can help to improve the ecological compatibility of cement clinker. This technology is based on biodegradable and cost-effective minerals and additives as well as a special complex admixture. The new method has the potential to increase the compressive strength of the innovative binder, improve the durability of biodegradable and cost-effective binder-based materials, and, at the same time, make use of inexpensive biodegradable minerals with industrial by-products and carbon quantum dots. This improvement leads to a reduction in energy consumption per unit of cement produced. Higher strength, better durability, reduction in pollution at the clinker production stage, and decrease in the landfill area occupied by industrial by-products are the ecological advantages of this biodegradable and cost-effective binder.

To achieve this objective, the authors of the book have endeavored to synthesize the findings of the existing literature and their research, which have been presented at international conferences and published in scientific journals, into a coherent and comprehensive account. The subject matter is addressed systematically and comprehensively, with each chapter of the book capable of being read and understood in isolation. However, the chapters are also structured in a way that ensures complementarity and avoids unnecessary repetitions, facilitating a holistic understanding of the subject. The book employs a pedagogical style, wherein the sequence of presentation of various sections facilitates comprehension of the barriers to achieving biodegradability and cost-effectiveness as sustainable attributes.

It is this author's opinion that the book has the quality and scope to become a leading reference in the literature concerning the general issue of converting an already efficient material from a technical performance point of view to an all-around optimal biodegradable and cost-effective construction material. This would be a significant advance, given that such a material has not been available in the last 150 years. Without economic and environmental benefits, progress is impossible. However, initiatives such as this book contribute to reducing the negative consequences of providing people around the globe with decent housing and infrastructure.

I wish to thank the chapter authors as well as Jenny Stanford Publishing for their contribution to the enhancement of the knowledge on the biodegradable and cost-effective binder.

Prof. Dr. Mehmet Serkan Kırgız
Academia of ACMCEN, USA

Chapter 1

Starch-Based Compound for Eco-friendly and Sustainable Construction Adhesive and Binder System

Mehmet Serkan Kirgiz
Academia of ACMCEN, Beaverton, OR 97075, USA
nakres42@yahoo.com

There is a need for a sustainable binder source because of the surplus CO_2 emission in manufacturing and the lack of raw natural resources for cement binders. Moreover, realizing a sustainable build environment and the United Nation's Sustainable Development Goals is the need of the hour. For this purpose, this chapter presents a number of current approaches for making sustainable and biodegradable binders and their composites, such as resin starch composites, coRncrete, Portland cement–starch fiber boards, and starch composites, in terms of the mixing approach of biodegradable binders, structure of starch binders, unit mass, thermal conductivity, compressive strength, and scanning electron microscopy (SEM)

Handbook of Innovative Adhesive Technology: Biodegradable and Cost-Effective Binder for Built Environment and Engineering Sciences
Edited by Mehmet Serkan Kirgiz

ISBN 978-981-5129-51-9 (Hardcover), 978-1-003-59892-3 (eBook)
www.jennystanford.com

analysis. The greatest density was found to be 1405 kg/m^3 in eco-friendly composites made of Portland cement (28.5%), starch (28.5%), water (28.5%), and jute fiber (14.2%) after 28 days. Besides, the density of the eco-friendly fiber board varied from 800 kg/m^3 to 1600 kg/m^3 in the study. By heating coRncrete at 200°C and using 250.8 kJ of energy, a compressive strength of 26.6 MPa was achieved.

Additionally, in the broad range of unit mass from 800 kg/m^3 to 1600 kg/m^3, the thermal conductivity differs from 0.49 W/mK to 0.66 W/mK. From SEM analysis, it was found that a 90:10 starch–polypropylene resin mixing ratio led to the most compact structure. Based on the results, the study achieved to develop several biodegradable binders and composites.

1.1 Introduction

Considering the crisis of energy and the lack of petroleum resources, discovering new and alternative binders has been in great demand since the beginning of 2022. The main aim of a new binder is to transform by-products into new materials such as artificial wood and for soil stabilization, making of meta-materials, and development of cement and concrete, and so on [1, 2]. Many researchers have studied starch-based binders because these are inexpensive natural polymers, a renewable source from agriculture, and easily biodegradable [3]. Nevertheless, the bonding of starch-based binders is not strong enough to glue materials together [4]. To overcome the lack of capacity, many studies have been performed, for instance, on composites with starch for thermal insulation [5–8], concrete with starch for viscosity modification [9–11], starch as modifier in asphalt-paving materials [12], vegetal concrete with different types of binders [13], and starch–hemp composite materials [8]. When the bonding feature of starch is examined, the following properties can be understood. Because starch is mixed with water in sufficient quantity, its powdered form makes a non-Newtonian liquid. Colloid starch shows the behavior of coagulation rapidly, which means that the colloid form of starch can endure deformation. When left undisturbed, it forms a thin liquid instead. Starch is a water-repellent powder in cold water. Besides, if starch powder is warmed in the presence of water, a transition of phase happens. The process of gelatinization is the transformation of starch into a gelatinous substance and this transformation results in the starch becoming insoluble in water [14]. The process of gelatinization is permanent, which involves swelling of the powder, melting of native crystalline structures, and solubility of molecule [15]. This chapter focuses on the application of starch-based compounds as additives for eco-friendly adhesive systems in new construction materials such as bio-based foam [16], bio-composites [17], hybrid timber with starch [18], Portland cement–starch composites [19], corn and wheat starch composites [20], and starch composite films [21].

1.2 Approaches of Mixing of Starch-Based Adhesive Systems for Eco-friendly Construction Materials

In the literature, many works used starch as an adhesive. In this section, the most current and up-to-date examples have been selected to better explain the approaches of mixing of starch-based adhesive systems for eco-friendly construction materials. All approaches have been summarized in Table 1.1. Fresh coRncrete was mixed with corn starch, sand, and potable water. The process of mixing was carried out for 2 min at 50 rpm through a mixer. Before wet-mixing, sand and starch were mixed to overcome agglomeration in wet-mixing. The process of mixing coRncrete is comprised of numerous steps. One of them is keeping the dry weight ratio of starch to sand constant at 20% (Table 1.1), and the water-to-starch ratio was 16.6%, which was determined from preliminary tests [22].

Starch-based bio-composites may be manufactured with lower density and better thermal conductivity compared to both Portland cement composites and resin composites. The density of starch-based composites can vary between 130 and 420 kg/m^3 [23–25]. The void in the internal structure of the composite results in low heat transfer, which changes from 0.048 W/mK to 0.11 W/mK. Nevertheless, it is strongly affected by the making process used [23, 24]. The ratio of hemp shiv for starch-based composites changes from 0.5 mm to

20 mm in Table 1.1 [24, 26]. Figure 1.1 shows the making of resin–starch foam. In Fig. 1.1, image (a) shows polypropylene resin 90% and wheat starch 10%; image (b) shows polypropylene resin 50% and wheat starch 50%; and image (c) shows polypropylene resin 20% and wheat starch 80%.

Table 1.1 Approaches of mixing of starch-based adhesive systems for eco-friendly construction materials

	Materials Ratio Used for Mixing (%)			
Examples of Starch-Based Composites	**Corn Starch/ Wheat Starch**	**Portland Cement/ Polypropylene Resin**	**Sand (0.125–0.25 mm)/ Hemp Shiv**	**Mixing and Production Process**
CoRncrete [22]	20/–	–/–	80/–	Mechanical mixing; plus tamping and heating at ≤200°C
Resin-starch foam [16]	–/50	–/50	–	Extrusion with heat at 190°C
Portland cement–starch fiber board [19]	28.57/–	28.57/–	–	Mechanical mixing
Starch-based bio-composites	60/60	–	–	Mechanical mixing and compressing at nearly 0.25 MPa. Then, drying at a temperature of 20°C ± 2°C and relative humidity of 50% ± 5%

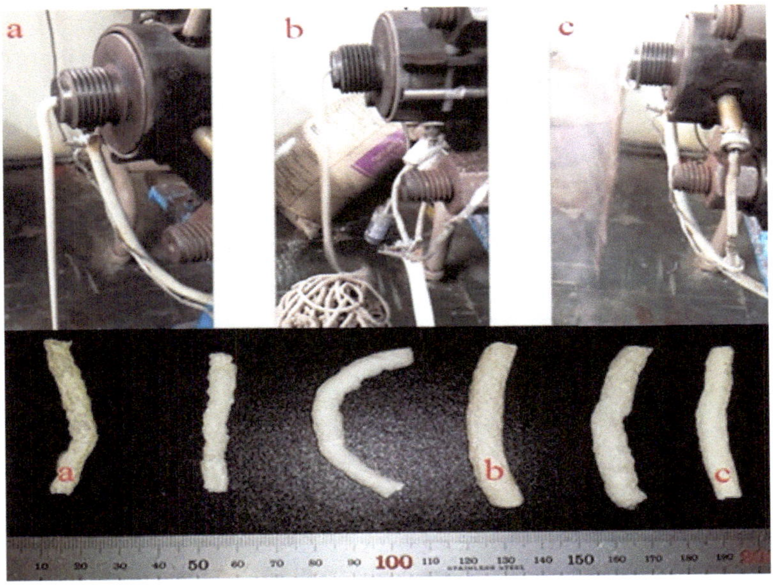

Figure 1.1 Making of resin–starch foam.

Figure 1.2 shows the making of Portland cement–starch fiber boards. For making these fiber boards, cement, starch, and jute fibers are combined. Raw farina is mixed without water to overcome agglomeration.

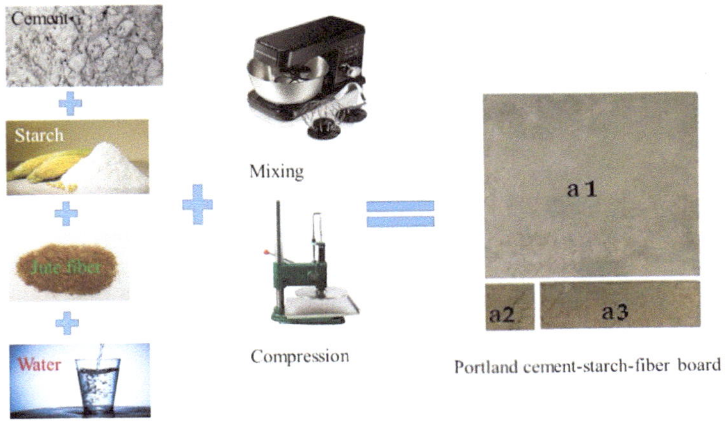

Figure 1.2 Making of Portland cement–starch fiber boards.

The mixture is then stirred with water for 10 min in a mixer at 30 rpm, unless otherwise stated. The fresh mixture is put into a mold to make a fiber board, which is compressed at 2.6 MPa. The resulting fiber board is cooled at room temperature. The dimensions of 300 × 300 mm^2, 50 × 50 mm^2, and 150 × 50 mm^2 are for the A1, A2, and A3 sizes, respectively. After the board becomes stiff, it is removed from the mold and dried at room temperature for 3 days (Fig. 1.2).

1.3 Structure and Features of Starch

The structure and features of starch have interested many researchers. With the advent of modern technology, researchers have gained a deeper understanding of the role of starch in composite systems and the subsequent development of serviceability. This section presents the most recent information on the structure and characteristics of starch, integrating approaches to the chemical composition, interactions, design, and function of starch-based materials. Starch is a plentiful, cheap, sustainable, and eco-friendly polymer. Maybe, it is the most plentiful polymer in nature, like cellulose, and can be derived mainly from the roots, stalks, and seeds of annual plants such as beans, oats, wheat, barley, and potato. Starch occurs as discrete granules whose dimension and form change according to plant resource, as can be understood from Table 1.2 [27]. Besides, starch is a polysaccharide, which is a polymer of the six-carbon sugar D-glucose, often known as the "building block" of starch.

Table 1.2 Physical properties of various starch granules

Type of Plant	Temperature Range of Gelatinization (°C)	Form of Granules	Dimension of Granules (nm)
Barley	51–60	Round or elliptical	20–25/2–6
Triticale	55–62	Round	19
Wheat	58–64	Lenticular or round	20–35/2–10

Type of Plant	Temperature Range of Gelatinization (°C)	Form of Granules	Dimension of Granules (nm)
Rye	57–70	Round or lenticular	28
Oats	53–59	Polyhedral	3–10
Potato	59–68	Oval	40
Corn	62–72	Round or polyhedral	15
Waxy maize	63–72	Round	15
Broad bean	64–67	Oval	30
Sorghum	68–78	Round	35
Rice	68–78	Polygonal	3–8

Figure 1.3 shows the structure of the monosaccharide D-glucose, which can be obtained either as an open chain or as a ring. Starch is composed of a unit of D-glucose, known as homoglucan or glucopyranose, and two major biomacromolecules, such as amylose and amylopectin.

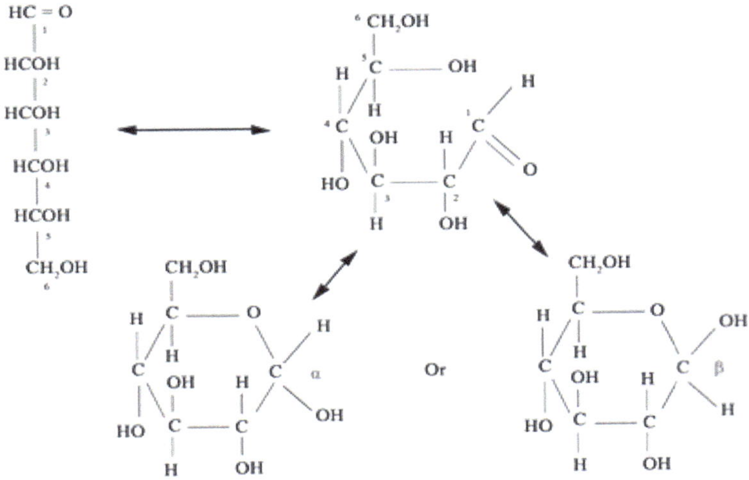

Figure 1.3 Structure of the monosaccharide D-glucose.

1.3.1 Amylose and Amylopectin of Starch

1.3.1.1 Amylose

Figure 1.4 shows the chemical structures of amylose and amylopectin. Amylose is a relatively lengthy and straight polymer consisting of D-glucopyranose of α-1.4-linking. The molecular weight of amylase is between 1×10^5 Da and 1×10^6 Da, and its degree of polymerization (DP) is between 324 and 4920, with approximately 9–20 points of branch, which is equivalent to chains 3–11 as per molecule [28, 29].

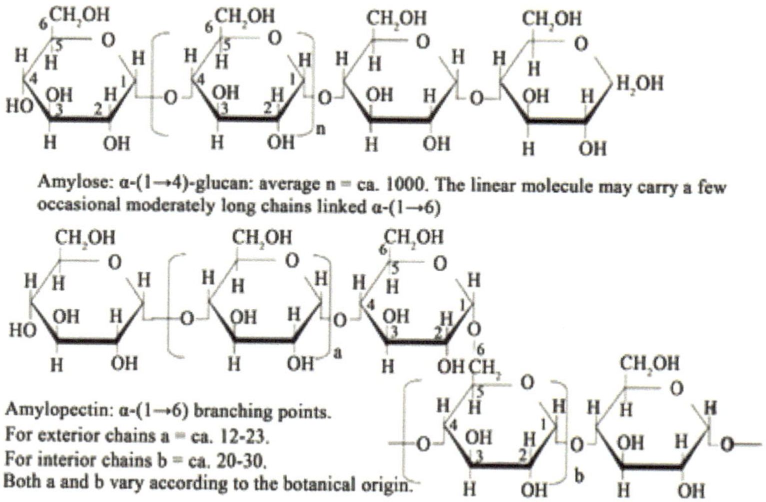

Figure 1.4 Chemical structures of amylose and amylopectin.

However, amylose has the capability to form a gel when starch powder is cooked with water. This feature is evident in the characteristics of certain amylose in starch [30, 31]. In most cases, gel-forming starches are employed, including corn starch, wheat starch, rice starch, and a hybrid corn plant that has high amylose corn starch isolated from it. The formation of gel results primarily in the degeneration of polymer solubilized when it is cooked and can happen rapidly along with the straight polymer of amylase [32–34].

1.3.1.2 Amylopectin

The scientific literature indicates that amylopectin molecules exhibit a variety of chemical structures and sizes. These include helical forms, chain-like structures, and patterns of clusters. Amylopectin is the primary molecule in many types of starch and is a branched polymer that is larger than amylose [35]. Amylopectin consists of the glucose part of the α-1.4-linking, which is connected by the branch point of the α-1.6-linking. A smaller chain of amylopectin has an average DP of approximately 15, while that of a larger chain of amylopectin is around 45 [29].

1.4 Other Compounds in Starch Powder

There are other compounds in starch, such as proteins, lipids, moisture, ash, minerals, and salts [36, 37], which are available in minute quantities [38, 39].

1.4.1 Different Types of Starch and Their Ratio of Amylose to Amylopectin

The amylose-to-amylopectin ratio within a given type of starch is an important point to consider with respect to starch functionality as adhesives. The contents of amylose and amylopectin in various starches are shown in Table 1.3 [27, 40].

Table 1.3 Contents of amylose and amylopectin in various starches

Type of Starch	Content of Amylose (%)	Content of Amylopectin (%)
Dent corn	25	75
Waxy corn	<1	>99
Tapioca	17	83
Potato	20	80
High-amylose corn	55–70 (or higher)	45–30 (or lower)
Wheat	25	75
Rice	19	81
Barley	22	—
Triticale	23–24	—
Rye	27	—
Oats	23–24	—
Broad bean	24	—
Sorghum	25	—

1.4.2 Eco-friendly Construction Materials with Starch

Starch is used as a binder in the manufacturing of artificial wood, thermal insulation materials, grout, and stucco. It can be either a single binder or a hybrid binder, such as Portland cement–starch binder, resin–starch binder, and lime–starch binder. Researchers have determined not only the physical properties but also the mechanical and thermal properties of eco-friendly construction materials made of starch. This subsection discusses the properties of eco-friendly construction materials containing starch.

1.4.3 Unit Mass

Considering the combination of only sand and starch, the density of eco-friendly materials is dependent on the content of water used. This might affect the penetration depth of a liquid after hardening. Materials such as coRncrete have been studied to understand the interaction between starch binder and sand and water. Since a starch binder is saturated with water, its shear strength increases linearly with an increase in the content of water. Besides, this results in void among sand and starch binder. There is a thin line that provides a balance for starch-based binders in construction materials. This is also the density-penetration depth parameter. Since the water content is between 0 and 10%, a reduction in dry density has been determined for coRncrete mixture. Nevertheless, the value of dry unit mass is significantly lower than that in other eco-friendly construction materials containing starch. The penetration depth of a liquid is in the same range of new materials. Because dry starch has no cohesive force, starch powder is free to transferring. An increase in water content, the powder of starch is much adhere than that of dry powder of starch, which may become cohesive. With an increase in water content from 10% to 14%, the cohesive strength decreases as per the increase in the depth of penetration. As the water content reaches 17% to 20%, the penetration is related to time. The optimum water content is 18% for coRncrete as researchers have obtained the greatest dry density of up to 1600 kg/cm^3. Above the content of water, fresh coRncrete does not display any force of shear [22]. In another work on starch-based foam, a combination of polypropylene resin (50%) and wheat starch (50%) showed the highest density. The density is dependent on the ratio of expansion and the ratio of foaming, which was 0.021 g/cm^3. It is considered that foaming of high expansion happened because the density of polypropylene resin is usually around 0.9 g/cm^3. Moreover, once the mixing ratio of resin to wheat starch was fixed at 50% and the content of water was free, the ratio of high expansion was 62 times greater than that of normal expansion, the ratio of foaming was 98.8%, and the density of the combination of polypropylene resin (50%) and wheat starch (50%) was 0.017 g/cm^3. This can be because there is a ratio of mixing and optimum content of water between the resin and wheat starch to activate as suitable foam in cushioning material. It is also necessary to go up the viscosity and to activate gelatinization of the starch by content of water to augment the ratio of volume expansion. The supramolecular assembly of starch is weakened due to the optimization of the proportion of starch, and the water enters and expands. Once water is vaporized because of high temperature, independent voids occur to form cells [16]. Unit mass is an indicator for the biodegradation of foam containing waxy corn starch and polylactic acid. When the biodegradation of foam occurs at a high rate, it may lead to such problems as circulation of product and commercialization. The degradability of waxy starch and polylactic acid was greater with a higher content of starch in eco-friendly cushioning composites. The results of the sample containing waxy corn starch (90%) and polylactic acid (10%) verified that the changing in the mixing ratio of the material can control degradability. Therefore, for the circulation of product and commercialization, degradability can be managed according to the time of circulation from the use of the material to its disposal by end users. In a previous study, Kozlowski *et al.* reported that cracks in composites made of starch and polylactic acid occurred initially due to the dispersion of starch [41]. Figure 1.5 shows density loss in eco-friendly composites, including waxy corn starch (90%) and polylactic acid (10%) [16]. Figure 1.6 shows the densities of various eco-friendly composites made of Portland cement, starch, water, and jute fibers.

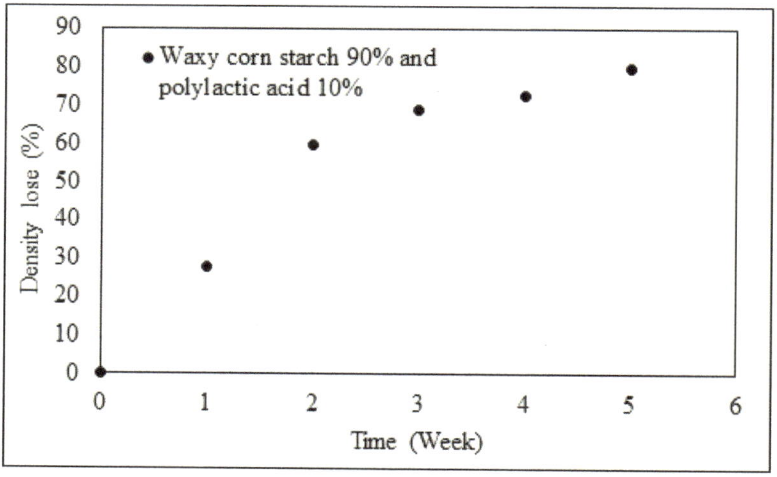

Figure 1.5 Density loss in eco-friendly composites, including waxy corn starch (90%) and polylactic acid (10%).

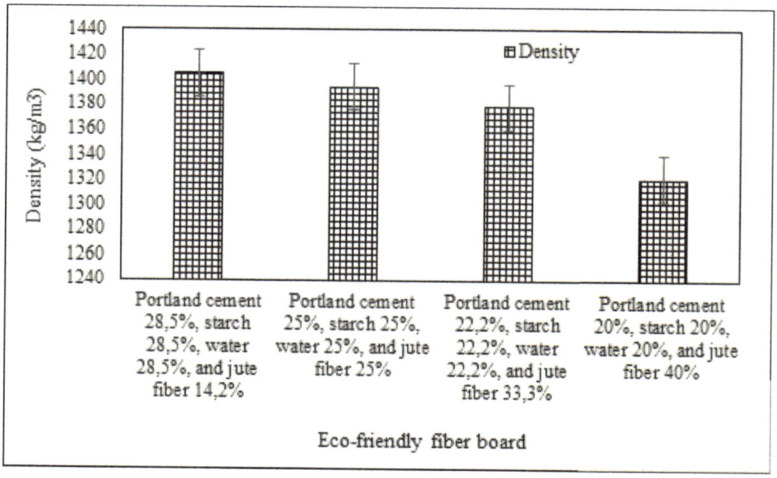

Figure 1.6 Densities of various eco-friendly composites made of Portland cement, starch, water, and jute fibers.

The design of foam containing waxy corn starch and polylactic acid is more eco-friendly and cheaper than that of foam made of resin products from petrochemical sources in the work, and it can be decomposed by organisms in nature, as verified by degradability measurement in density loss in Figs. 1.5 and 1.6 [16]. Figure 1.7 shows the comparative densities of the composites.

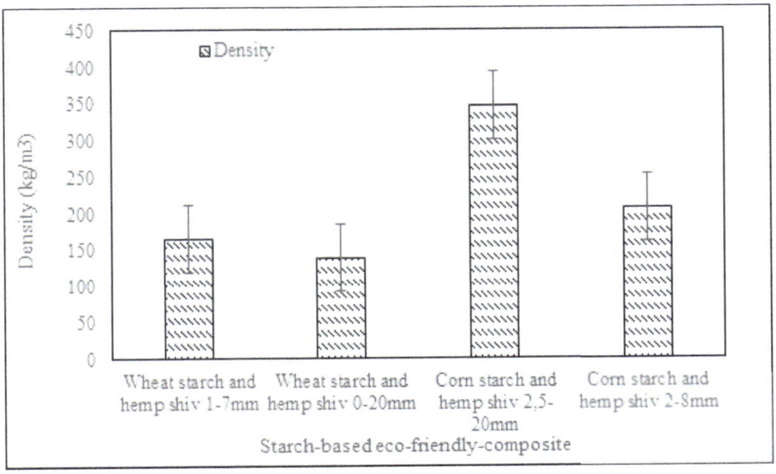

Figure 1.7 Density of starch-based eco-friendly composites.

It was found that the greatest density was 1405 kg/m^3 in eco-friendly composites made of Portland cement (28.5%), starch (28.5%), water (28.5%), and jute fiber (14.2%) after 28 days. Besides, the density of eco-friendly fiber boards varied from 1321 kg/m^3 to 1405 kg/m^3, and it was reported that the higher the proportion of jute fibers added, the lesser the density achieved.

The last study presented here on starch-based eco-friendly composites proves that these can be manufactured with low density. The density of starch-based eco-friendly composites can vary between 130 kg/m^3 and 420 kg/m^3 [23–25]. For eco-friendly composites with a shorter length of hemp shiv, the density is between 163 kg/m^3 and 169 kg/m^3, while for composites with longer hemp shiv of length 0–20 mm, the density linearly increases from 168 kg/m^3 to 174 kg/m^3. The ratio of water to starch binder in the composite mostly varies from 0.4 to 0.6. The properties of eco-friendly composites are mostly affected by various ratios of hemp shiv to starch binder, which lie in the broad range of 2 to 14. Moreover, the increase in the ratio of hemp shiv to starch binder increases the absorption of water by up to 12% with 56% humidity in the composites [19].

1.4.4 Thermal Conductivity

Figure 1.8 shows the thermal conductivity of Portland cement–starch fiber boards, Portland cement–starch boards, and jute fibers. With the addition of jute, the thermal conductivity decreases in the Portland cement–starch fiber boards. In Portland cement–starch fiber boards with and without jute fibers, no important difference has been observed, and the thermal conductivity is similar to those prepared with jute fibers without any binder. Consequently, in making Portland cement–starch fiber boards, it has been found that they are good thermal insulators, having a lower unit weight than jute fibers. Portland cement–starch fiber boards with no jute fibers have higher unit weight and greater thermal conductivity. Therefore, it has the worst qualification as a material for thermal insulation [19].

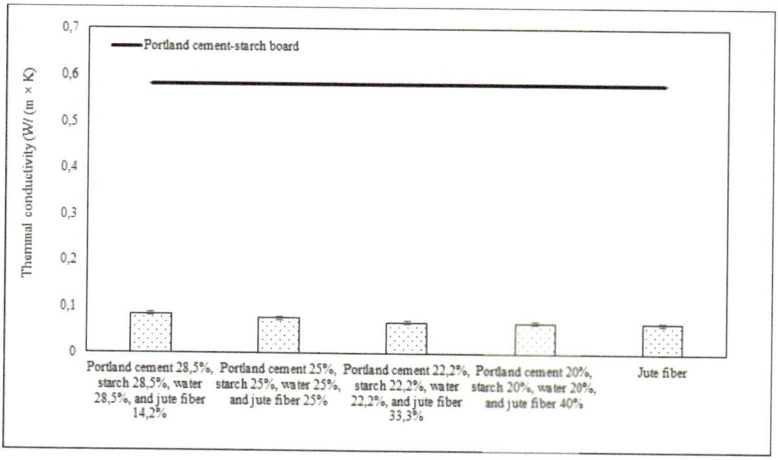

Figure 1.8 Thermal conductivity of Portland cement–starch fiber boards, Portland cement–starch boards, and jute fibers.

The porosity and thermal conductivity are inversely proportional to each other between the properties of starch-based materials. The packing of a supplement into the phase leads to porosity. In addition, a material with heat-insulating property differs from its unit weight inversely. For instance, the property of heat insulation increases because the unit weight decreases. However, the point is that while the thermal conductivity is kept nearly constant, the unit weight must be lower than the high unit weight provided by the high thermal conductivity. It was known that the unit weight of the starch-based phase was higher than the unit weight made of jute fibers only. The porosity of the matrix increases with the addition of fibers. The loss of thermal conductivity is due to an increase in porosity because of the content quantity of supplement fibers, other binders, aggregates, and by-products. By blending lower than 33wt.% jute fiber, it seems that the porosity does not increase. Thus, the thermal conductivity does not increase as well. To develop the thermal conductivity of Portland cement–starch fiber boards, it is required to keep the ratio between the length and the diameter of the fiber constant, achieve the ideal content of water, and optimize the interface zone between the binder phase and jute fibers. Portland cement–starch fiber boards made of cement, sand, and coconut fibers had a unit weight of 1586.8 kg/m^3 and a thermal conductivity of 0.6 W/mK. With the blending of 0.75% hemp fibers and a reduction in coarse aggregates (10% by volume of the binder), the achieved unit weight and thermal conductivity were 2192 kg/m^3 and 1.7 W/mK, respectively. Therefore, Portland cement–starch fiber boards show greater thermal conductivity and lesser unit weight than those made of cement and jute fibers only, which makes Portland cement–starch fiber boards effective insulators for construction [19]. Figure 1.9 shows the thermal conductivities of wheat starch–hemp boards, corn starch–hemp boards, gypsum boards with cork aggregates, and polymer resin–metakaolin boards with wheat straw aggregates.

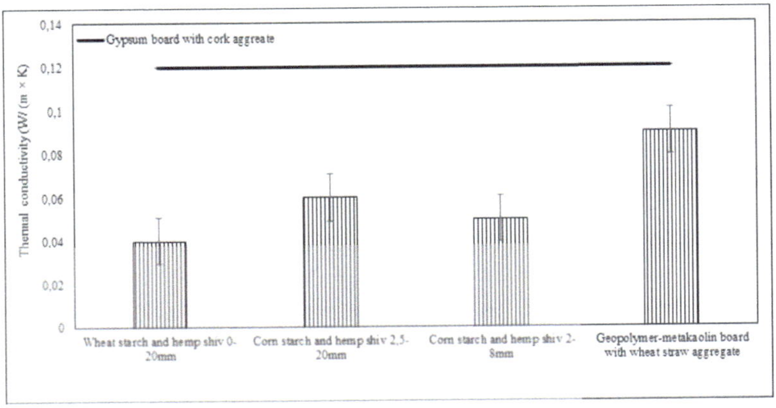

Figure 1.9 Thermal conductivities of wheat starch–hemp boards, corn starch–hemp boards, gypsum boards with cork aggregates, and geopolymer–metakaolin boards with wheat straw aggregates.

Porosity describes a lower thermal conductivity, which differs from 0.04 W/mK to 0.1 W/mK. Besides, it is strongly affected by the forming factor and the production method used. The portion of hemp shiv used for starch-based bio-composites differs from 0.5 mm to 20 mm. For the prepared bio-composites, a compression force of nearly 0.25 MPa is usually applied. Then, the bio-composite is dried at a temperature of 20 ± 2°C and a relative humidity of 50 ± 5%. Bio-composites with a density of 184–456 kg/m^3 have obtained a compressive strength of 4–71 kPa and a thermal conductivity (λ) of 0.05–0.08 W/mK, which have been made with a straw–gypsum binder ratio between 0.2 and 0.4. The utilization of bio-fillers such as corks leads to obtaining gypsum bio-based composites with a unit weight of 800 kg/m^{-3} and a relatively high thermal conductivity of 0.3 W/mK. It has been shown that because 2–6% of wheat straw fiber for the gypsum mass was used in the bio-composite material, there was a suggestion to use optimum content, such as not more than 3%. In the broad range of apparent unit weight from 800 kg/m^3 to 980 kg/m^3, compressive strength and thermal conductivity differ from 2 MPa to 2.5 MPa and from 0.49 W/mK to 0.66 W/mK, respectively [20].

1.4.5 Compressive Strength

Figure 1.10 shows the compressive strength of coRncrete with natural aggregates, manufactured in either a microwave oven or a classical oven. Many factors affect the compressive strength of starch-based materials, such as the water content, grain size of supplements, and the heating period in a classical or microwave oven.

As understood from Fig. 1.10, the compressive strength increases as the water content increases up to 15% in a microwave oven–processed coRncrete. Of all coRncrete heated in an oven, there is a certain decrease in the compressive strength because it contains very low water content. In all occasions of compression, brittle fracture was seen. With an increase in water content to 15%, the compressive strength reached 1.3 MPa. The optimum content of water for fresh coRncrete is 16.6% since the compressive strength of coRncrete increases by that percent. With an optimum content of water, coRncrete achieved the maximum compressive strength of

13.7 MPa. An increase in the content of water to 20% starts reducing the strength of compression. In the case of heating in a microwave, a compressive strength of 1.8 MPa was seen in coRncrete with 10% content water. With a rise in the content of water to 15%, the strength rises rapidly, and 26.7 MPa has been recorded for compressive strength, which is the known maximum strength of compression. On exceeding the optimum content of water, 21.6 MPa was managed for the compressive strength of coRncrete, which is the known lowest strength. At the optimum content of water, the strength of compression starts to reduce, and 15.3 MPa has been managed for coRncrete at 18.3% water content. Because it was prepared with 20% water content, it deformed easily. Therefore, it was unsuitable for testing compressive strength. This results because of heating coRncrete rapidly and approaching the boundary condition, that is, the temperature of gelatinization, in due course. Figure 1.11 shows the compressive strengths of particleboards manufactured with wheat-based starch or corn-based starch.

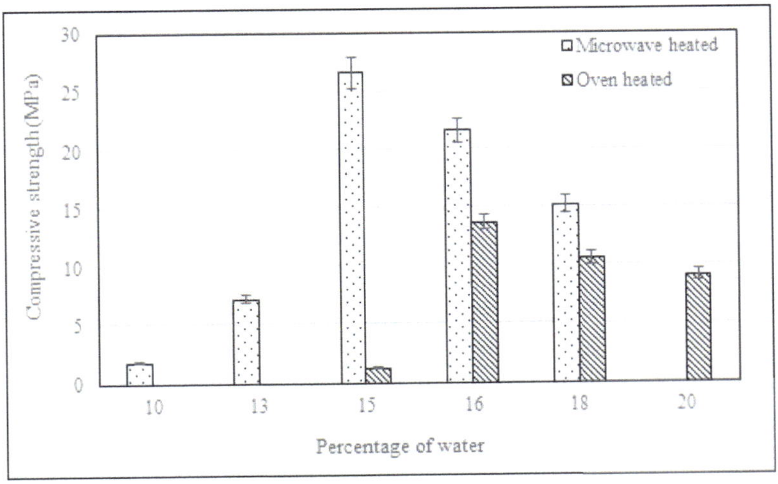

Figure 1.10 Compressive strength of coRncrete with natural aggregates, manufactured in either a microwave oven or a classical oven.

The grain size of supplements has an important impact on the compressive strength of starch-based, eco-friendly, and sustainable construction adhesive and binder systems. As understood from Fig. 1.11, compressive strength increases with an increase in the grain size of supplements. The situation is different for coRncrete because its compressive strength is influenced by the content of water and heating as well as the grain size of aggregates. It has been shown that the compression strength of coRncrete made of 0.5–1.0 mm grains is greater than that of coRncrete made of 0.25–0.5 mm grains. This probably results in the burning of corn starch, which contains 0.25–0.5 mm grains. The effect of the grain size of supplements on the strength of starch-based, eco-friendly, and sustainable construction adhesive and binder systems can be better understood by the following explanation: With a rise in the size of grains, the specific surface area of the supplement increases and the quantity of bonding decreases intergranularly, which reduces the interfacial zone between the supplement's grains and the starch binder. Starch-based, eco-friendly, and sustainable construction adhesive and binder systems such as particleboards show an opposite trend of gaining compressive strength along with a higher grain size of supplements. Therefore, a lower force of bonding breaks the specimen with no applied compression force, which lessens the interfacial zone of bonding between the grains of the supplement and the starch binder. In starch-based, eco-friendly, and sustainable construction adhesive and binder systems, there is another factor known as the period of heating in a classical oven or microwave oven. The effect has been described as the process of gelatinization and the bonding between the supplement's grains and the starch binder. An important loss of moisture has also been observed during this period of heating. After the heating, the gain in compressive strength is low and the moisture content is high in hardened coRncrete, which is another starch-based, eco-friendly, and sustainable construction adhesive and binder material. On heating coRncrete at 200°C, which is 250.8 kJ of energy, 26.6 MPa was managed for compressive strength. On heating higher, the loss of moisture occurred tough since the bonding is firmly kept by starch binder gelatinized. It is significant to mark that apart from the factors discussed in the chapter: The period and degree of heating, the size of supplement grain, the content of water, and the loss of moisture are actively related to the gain of compressive strength in starch-based, eco-friendly, and sustainable construction adhesive and binder systems.

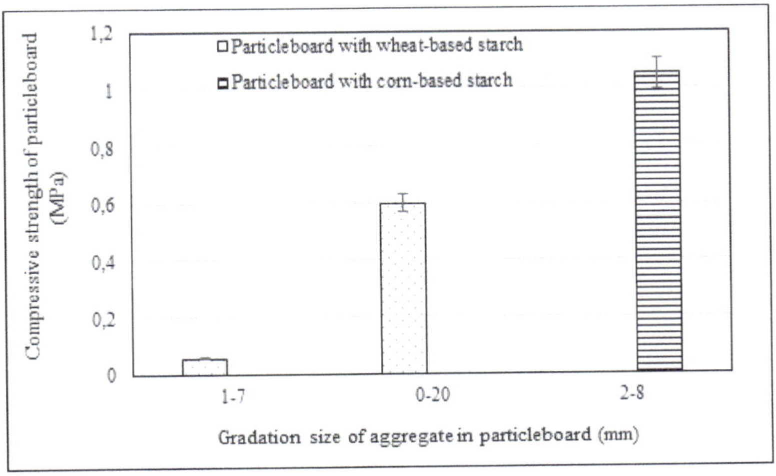

Figure 1.11 Compressive strengths of particleboards manufactured with wheat-based starch or corn-based starch.

1.4.6 SEM Analysis

Environmental SEM images obtained the morphology of starch-based, eco-friendly, and sustainable construction adhesive and binder systems for analysis at ×60, ×120, ×1500, and ×4000 magnifications. Figure 1.12 shows resin–starch foam containing 100% starch condition. The cell wall is well formatted, and the foaming of high-magnification was persuaded.

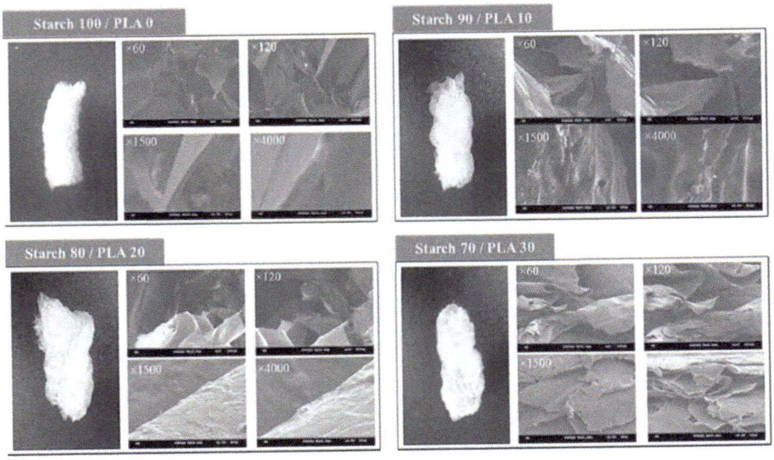

Figure 1.12 Resin–starch foam containing 100% starch condition.

However, the foam is brittle. Therefore, it is unsuitable for gaining compressive strength and for cushioning in packing materials. The cell wall is thin, as in vinyl, since starch and polypropylene resin are mixed and foamed. The 80:20 mixing ratio of starch binder to polypropylene resin, which approaches 90:10, leads to a relatively compact structure. Comparing those content with the mixing ratio of 70:30, the thickness of cell wall rises. Figure 1.13 demonstrates the morphology of the cross section of a cushioning material made by mixing 50% polypropylene resin and 50% wheat starch binder, while the water content was 20%. The cell wall demonstrated a vinyl shape angled, which is unlike the cell wall of usual polymer resin foam. Moreover, the cell wall was comparatively thin. The morphology of cell confirmed that the process of foaming was properly fulfilled.

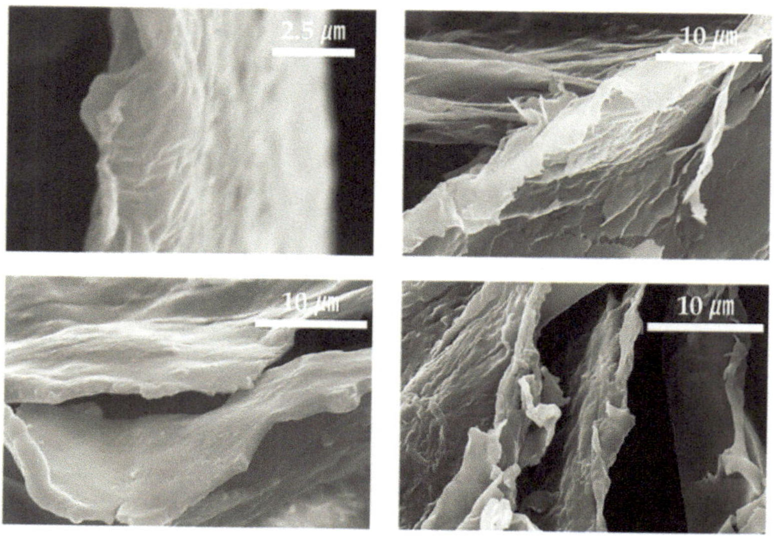

Figure 1.13 Morphology of cross section of a cushioning material made by mixing 50% polypropylene resin and 50% wheat starch binder, with a 20% water content.

Composites of resin–starch foam showed various ratios of volume expansion with respect to the ratios of mixing materials. The content of 50% polypropylene resin and 50% wheat starch binder has shown the greatest quantity and has been acknowledged to be applicable as a cushioning material. In addition to this knowledge, the tissue is very different in the mixing ratio of polypropylene resin (50%) and wheat starch binder (50%). Because the content of polypropylene resin increases, the resin–starch foam is tough. In addition, because the content of wheat starch binder increases, the resin–starch foam becomes directly brittle. This meant that the developed the foam of resin-starch foam material gets lesser force when it failures. There is no difference between conventional cushioning and resin–starch foam cushioning as the impulse of the commercial product is 2.810 kgm/s, while that of the resin–starch foam cushioning was recorded as 2.530 kgm/s in the chapter. This means that resin–starch foam can be used as a commercial cushioning material.

References

1. Imam, S. H., Gordon, S. H., Mao, L., and Chen, L. Environmentally friendly wood adhesive from a renewable plant polymer: Characteristics and optimization. *Polymer Degradation and Stability*, **73**(3), 2001: 529–533. https://doi.org/10.1016/s0141-3910(01)00114-8.
2. Pizzi, A. Bioadhesives for wood and fibres. *Reviews of Adhesion and Adhesives,* **1**(1), 2013: 88–113. https://doi.org/10.7569/raa.2013.097303.
3. Kennedy, H. G. Starch- and dextrin-based adhesives. *American Chemical Society EBooks*, 1989, 326–336. https://doi.org/10.1021/bk-1989-0385.ch023.
4. Xu, W., J. Shi, and S. Wang. Study on heat aging properties of starch based aqueous polymer isocyanate adhesive for wood. *Advanced Materials Research*, 2014. https://doi.org/10.4028/www.scientific.net/amr.933.138.
5. Alvarez, V. A., A. Vázquez, and C. R. Bernal. Fracture behavior of sisal fiber-reinforced starch-based composites. *Polymer Composites,* **26**(3), 2005: 316–323. https://doi.org/10.1002/pc.20103.
6. Balčiūnas, G., S. Vėjelis, S. Vaitkus, and A. Kairytė. Physical properties and structure of composite made by using hemp hurds and different binding materials. *Procedia Engineering*, **57**, 2013: 159–166. https://doi.org/10.1016/j.proeng.2013.04.023.
7. Sandrine, U. B., V. Isabelle, M. P. Hoang, and M. Chadi. Influence of chemical modification on hemp–starch concrete. *Construction and Building Materials,* **81**, 2015: 208–215. https://doi.org/10.1016/j.conbuildmat.2015.02.045.

8. Le, A., A. Gacoin, A. Li, T. H. Mai, M. Rebay, and Y. Delmas. Experimental investigation on the mechanical performance of starch–hemp composite materials. *Construction and Building Materials*, **61**, 2014: 106–113. https://doi.org/10.1016/j.conbuildmat.2014.01.084.
9. Cappellari, M., D. Anne, and M. Chaouche. Influence of organic thickening admixtures on the rheological properties of mortars: Relationship with water-retention. *Construction and Building Materials*, **38**, 2013: 950–961. https://doi.org/10.1016/j.conbuildmat.2012.09.055.
10. Izaguirre, A., J. Lanas, and J. I. Alvarez. Behaviour of a starch as a viscosity modifier for aerial lime-based mortars. *Carbohydrate Polymers*, **80**(1), 2010: 222–228, https://doi.org/10.1016/j.carbpol.2009.11.010.
11. Rols, S., J. Ambroise, and J. Pera. Effects of different viscosity agents on the properties of self-leveling concrete. *Cement and Concrete Research*, **29**(2), 1999: 261–266. https://doi.org/10.1016/s0008-8846(98)00095-7.
12. Al-Hadidy, A. I., T. Yi-Qiu, and A. T. Hameed. Starch as a modifier for asphalt paving materials. *Construction and Building Materials*, **25**(1), 2011: 14–20. https://doi.org/10.1016/j.conbuildmat.2010.06.062.
13. Lagouin, M., C. Magniont, P. Sénéchal, P. Moonen, J.-E. Aubert, and A. Laborel-Préneron. Influence of types of binder and plant aggregates on hygrothermal and mechanical properties of vegetal concretes. *Construction and Building Materials*, **222**, 2019: 852–871. https://doi.org/10.1016/j.conbuildmat.2019.06.004.
14. Donovan, J. E. Phase transitions of the starch–water system. *Biopolymers*, **18**(2), 1979: 263–275. https://doi.org/10.1002/bip.1979.360180204.
15. Xie, F., P. Liu, and L. Yu. Processing of plasticized starch-based materials: State of the art and perspectives. İn *Starch Polymers: From Genetic Engineering to Green Applications*, Elsevier B.V. (2014): pp. 257–289. https://doi.org/10.1016/B978-0-444-53730-0.00024-5.
16. Sohn, J. H., H. J. Kim, and S. W. Cha. Bio-based foamed cushioning materials using polypropylene and wheat bran. *Sustainability*, **11**(6), 2019: 1670. https://doi.org/10.3390/su11061670.
17. Curto, D., A. Guercio, and V. Franzitta. Investigation on a bio-composite material as acoustic absorber and thermal insulation. *Energies*, **13**(14), 2020: 3699. https://doi.org/10.3390/en13143699.
18. Wu, T.-C., M. Gong, and J. Xiao. Preliminary sensitivity study on an life cycle assessment (LCA) tool via assessing a hybrid timber building. *Journal of Bioresources and Bioproducts*, **5**(2), 2020: 108–113. https://doi.org/10.1016/j.jobab.2020.04.004.

19. Ferrandez-Garcia, M. T., C. E. Ferrandez-Garcia, T. Garcia-Ortuño, A. Ferrandez-Garcia, and M. Ferrandez-Villena. Study of waste jute fibre panels (*Corchorus capsularis* L.) agglomerated with Portland cement and starch. *Polymers*, **12**(3), 2020: 599. https://doi.org/10.3390/polym12030599.

20. Bumanis, G., L. Vitola, I. Pundiene, M. Sinka, and D. Bajare. Gypsum, geopolymers, and starch—alternative binders for bio-based building materials: A review and life-cycle assessment. *Sustainability*, **12**(14), 2020: 5666. https://doi.org/10.3390/su12145666.

21. Huang, L., H. Zhao, T. Yi, M.-H. Qi, H. Xu, Q. Mo, C. Huang, S. Wang, and Y. Liu. Preparation and properties of cassava residue cellulose nanofibril/cassava starch composite films. *Nanomaterials*, **10**(4), 2020: 755. https://doi.org/10.3390/nano10040755.

22. Kulshreshtha, Y., E. Schlangen, H. M. Jonkers, P. J. Vardon, and L. A. van Paassen. CoRncrete: A corn starch based building material. *Construction and Building Materials*, **154**, 2017: 411–423. https://doi.org/10.1016/j.conbuildmat.2017.07.184.

23. Balčiūnas, G., S. Vejelis, L. Lekunaite, and A. Kremensas. Assessment of structure influence on thermal conductivity of hemp shives composite. *Environmental Engineering and Management Journal*, **15**(3), 2016: 699–705. https://doi.org/10.30638/eemj.2016.075.

24. Bourdot, A., T. Moussa, A. Gacoin, C. Maalouf, P. Vázquez, C. Thomachot-Schneider, *et al.* Characterization of a hemp-based agro-material: Influence of starch ratio and hemp shive size on physical, mechanical, and hygrothermal properties. *Energy and Buildings*, **153**, 2017: 501–512. https://doi.org/10.1016/j.enbuild.2017.08.022.

25. Le, A., A. Gacoin, A. Li, T. H. Mai, and N. El Wakil. Influence of various starch/hemp mixtures on mechanical and acoustical behavior of starch-hemp composite materials. *Composites Part B-Engineering*, **75**, 2015: 201–211. https://doi.org/10.1016/j.compositesb.2015.01.038.

26. Curvelo, A. A. S., A. J. F. De Carvalho, and J. A. M. Agnelli. Thermoplastic starchcellulosic fibers composites: Preliminary results. *Carbohydrate Polymers*, **45**(2), 2001: 183–188. https://doi.org/10.1016/S0144-8617(00)00314-3.

27. Lineback, D. R. Current concepts of starch structure and its impact on properties. *Journal of the Japanese Society of Starch Science*, **33**(1), 1986: 80–88. https://doi.org/10.5458/jag1972.33.80.

28. Buléon, A., P. Colonna, V. Planchot, and S. G. Ball. Starch granules: Structure and biosynthesis. *International Journal of Biological*

Macromolecules, **23**(2), 1998: 85–112. https://doi.org/10.1016/s0141-8130(98)00040-3.

29. Yoshimoto, Y., T. Takenouchi, and Y. Takeda. Molecular structure and some physicochemical properties of waxy and low-amylose barley starches. *Carbohydrate Polymers*, **47**(2), 2002: 159–167. https://doi.org/10.1016/s0144-8617(01)00184-9.
30. Tester, R. F., and J. Karkalas. Starch. *Biopolymer*, **6**, 2002: 381–438.
31. Wang, L. Z., and P. J. White. Structure and properties of amylose, amylopectin, and intermediate materials of oat starches. *Cereal Chemistry*, **71**(3), 1994: 263–268.
32. Tester, R. F., and W. R. Morrison. Swelling and gelatinisation of cereal starches. I. Effects of amylopectin. Amylose and lipids. *Cereal Chemistry*, **67**, 1990a: 551–557.
33. Tester, R. F., and W. R. Morrison. Swelling and gelatinisation of cereal starches. II. Waxy rice starches. *Cereal Chemistry*, **67**, 1990b: 558–563.
34. Tester, R. F., and W. R. Morrison. Swelling and gelatinisation of cereal starches. III. Some properties of waxy and normal non-waxy barley starches. *Cereal Chemistry*, **69**, 1992: 645–658.
35. Takeda, Y., S. Shibahara, and I. Hanashiro. Examination of the structure of amylopectin molecules by fluorescent labeling. *Carbohydrate Research*, **338**(5), 2003: 471–475. https://doi.org/10.1016/s0008-6215(02)00488-3.
36. Greenwell, P., and J. D. Schofield. A starch granule protein associated with endosperm softness in wheat. *Cereal Chemistry*, **63**(4), 1986: 379–380. http://pascal-francis.inist.fr/vibad/index.php?action=getRecordDetail&idt=8785658.
37. Rayas-Duarte, P., S. F. Robinson, and T. P. Freeman. In situ location of a starch granule protein in durum wheat endosperm by immunocytochemistry. *Cereal Chemistry*, **72**(3), 1995: 269–274.
38. Acker, L. The role of starch lipids among cereal lipids, their composition and their importance for the baking properties of wheat flours. *Getreide, Mehl und Brot* (English title: Cereals, flour and bread), **36**, 1982: 291–295.
39. Biliaderis, C. G., C. A. Page, and T. J. Maurice. On the multiple melting transitions of starch/monoglyceride systems. *Food Chemistry*, **22**(4), 1986: 279–295. https://doi.org/10.1016/0308-8146(86)90086-5.
40. Gadhave, R. V., P. A. Mahanwar, and P. T. Gadekar. Starch-based adhesives for wood/wood composite bonding: Review. *Open Journal*

of Polymer Chemistry, **7**(2), 2017: 19–32. https://doi.org/10.4236/ojpchem.2017.72002.

41. Kozlowski, M., R. Masirek, E. Piorkowska, and M. Gazicki-Lipman. Biodegradable blends of poly(l-lactide) and starch. *Journal of Applied Polymer Science*, **105**(1), 2007: 269–277. https://doi.org/10.1002/app.26088.

Chapter 2

Self-Compacting Composite Containing Mine Tailing and Fortified by Nanocarbon Dot for Manufacturing of Cost-Effective and CO_2-Neutral Material in Construction and Infrastructure

Mehmet Serkan Kirgiz,[a] André G. D. S. Galdino,[b] Ayub Elahi,[c] and Atizaz Ali[c]
[a]*Academia of ACMCEN, Beaverton, OR 97075, USA*
[b]*Federal Institute of Education, Sciences and Technology of Espírito Santo, Av. Vitória, 1729, Jucutuquara, Vitória, ES, 29040-780, Brazil*
[c]*Department of Civil Engineering, UET Taxila, Pakistan*
nakres42@yahoo.com

This study aims to make environmentally friendly and sustainable self-compacting composite (SCC) using waste materials and nanocarbon dots. Mine tailings, which are byproducts of mining, were used as a partial replacement for ordinary Portland cement

Handbook of Innovative Adhesive Technology: Biodegradable and Cost-Effective Binder for Built Environment and Engineering Sciences
Edited by Mehmet Serkan Kirgiz

ISBN 978-981-5129-51-9 (Hardcover), 978-1-003-59892-3 (eBook)
www.jennystanford.com

(OPC) in SCC. To activate the waste material, it was pretreated and ground mechanically. A detailed experimental program was designed and executed for optimizing the SCC mix with nanocarbon dots. The results suggested that SCC could be made by incorporating mine tailing powder and nanocarbon dots. The use of mine tailing powder is cheaper than that of other supplementary cementitious materials (SCMs). As a result, 15% mine tailing powder along with nanocarbon dots increased the compressive strength by 17.7%, splitting tensile strength by 16.2%, and flexural strength by 22.2% when compared to the control SCC mix. In addition, the use of waste materials as a partial replacement of OPC in SCC can reduce the quantity of OPC and viscosity-modifying admixtures (VMA) to make the SCC favorable. As could be understood from the results, cost-effective, sustainable, and environmentally friendly SCC can be produced for the Sustainable Development Goals (SDGs) of the United Nations.

2.1 Introduction

Self-compacting composite is a kind of high-performance material that consolidates under its self-weight [1]. SCC presents several advantages over traditional concrete, including improved surface finish, reduced noise pollution, faster placement without mechanical compaction, and easier pumping. Therefore, it can be a favorable alternative to conventional concrete for various construction applications. SCMs such as fly ash, metakaolin, silica fume, and ground granulated blast furnace slag are used in concrete for various purposes. One of the main purposes of SCMs in concrete is reducing OPC consumption in construction applications. This reduction not only improves the environmental sustainability of the concrete production process by reducing the carbon footprint associated with OPC production, but also contributes to cost saving, making SCM usage an attractive option for concrete production as well as fulfilling the UN SDGs. Waste materials such as mine tailings are among the potential supplementary cementitious sources. A few research studies have highlighted the use of wastes from mines [2, 3]. It has been observed from the literature review that there is a significant lack of information on the use of mine tailings as SCMs in self-compacting concrete. To overcome this knowledge gap, the present study was conducted to evaluate the effect of the material together with nanocarbon dots in the making of green and sustainable self-compacting concrete. This study is, therefore, crucial to carry out cleaner production that can overcome the challenges faced by the construction industry in every region.

2.2 Materials and Methods

In this study, detailed experimental investigations were conducted to produce a greener and more sustainable version of SCC. Three natural waste powder materials were used as a partial replacement of OPC for optimizing the maximum possible replacement dosage.

2.2.1 Materials

Type 1 cement conforming to the ASTM standard C150-07 was used [4]. The fineness and specific gravity of OPC were 2735 cm^2/g and 3.11 g/cm^3, respectively. The specific gravity values for coarse aggregates and fine aggregates were 2.74 g/cm^3 and 2.64 g/cm^3, respectively. Siliceous mine tailings were taken from a quartz-based mine. A polycarboxylate-based high-range water reducer, known as Sika Viscocrete 20-HE, was used to make an admixture much more flowable.

2.2.2 Pretreatment to Improve the Effectiveness of SCM and Nanocarbon

For the mechanical activation of mine tailings and nanocarbon dots separately, a ball mill was used. In the ball mill, steel balls of five, eight, and fifteen diameters were used to homogenize the SCM stance [5, 6]. The 30 min cooling interval was used to avoid excessive temperature. To know the chemical composition of mine tailings and nanocarbon dots, wet chemical analysis was performed. To produce environmentally friendly SCC, 13 self-compacting concrete mixes were cast using different types of SCMs, percentages, and cement content. The treated SCM was used for making SCC mixes after 30 min of grinding. The mixes were divided into three series. Table 2.1 summarizes the SCC mixture proportions for 1 m^3.

Table 2.1 Mix design of SCC for 1 m^3

	Proportions of Mixture (kg/m^3)							
Mix ID	**Water-to-Binder Ratio**	**OPC**	**Fine Aggregate**	**Coarse Aggregate**	**Water**	**Mine Tailing Powder**	**Nano-carbon dots**	**SP**
S-0	0.37	505.00	844	783	187	0	0	7.58
SM-1	0.37	479.75	844	783	187	25.25	1.25	7.58
SM-2	0.37	429.25	844	783	187	75.75	1.25	7.58

The loss on ignition and chemical properties of blended and reference cement paste samples were measured as per the standard EN 196-2 (TSE 2002) [7], known as wet analysis. Cylinders of 150 mm diameter and 300 mm height were cast to determine the compressive strength as per ASTM C39/C39M-21 [8] at 3, 28, and 56 days of curing. Cylinders of the same size were cast to determine flexural and split tensile strengths as per ASTM C496/C496M-17 [9] at 28 days of curing. For each strength property, three samples were tested. The average was considered the strength property of the SCC.

2.3 Results and Discussion

2.3.1 Chemical Component

Table 2.2 provides the chemical composition of the mine tailing powder sample and nanocarbon dots after grinding.

Table 2.2 Chemical composition of mine tailing and nanocarbon dot samples

Oxide Composition	Sample of Mine Tailing (%)	Sample of Nanocarbon (%)
Silica (SiO_2)	86.8	—
Aluminum (Al_2O_3)	6.63	—
Iron (Fe_2O_3)	4.01	—
Magnesium (MgO)	0.24	—
Calcium (CaO)	0.64	—
Sodium (Na_2O)	0.32	—

Oxide Composition	Sample of Mine Tailing (%)	Sample of Nanocarbon (%)
Potassium (K_2O)	0.50	—
Titanium (TiO_2)	0.88	—
Carbon (C)	—	99.9
Loss on ignition (LOI)	0.10	0.01

If the percentage of $SiO_2 + Al_2O_3 + Fe_2O_3$ is equal to or more than 70, the material is classified as pozzolan as per ASTM C618-17 [10]. From Table 2.1, it can be observed that the mine tailing has the $SiO_2 + Al_2O_3 + Fe_2O_3$ percentage of 97.44 and the nanocarbon dots show the presence of unburned carbon in its inner structure. Therefore, nanocarbon dots are activators in the advance treatment of cement-based materials containing SCMs.

2.3.2 Compressive Strength

Figure 2.1 shows the compressive strength of various concrete mixtures made with different replacement levels of mine tailings. The control mixture of SCC has a compressive strength of 40 MPa at 28 days.

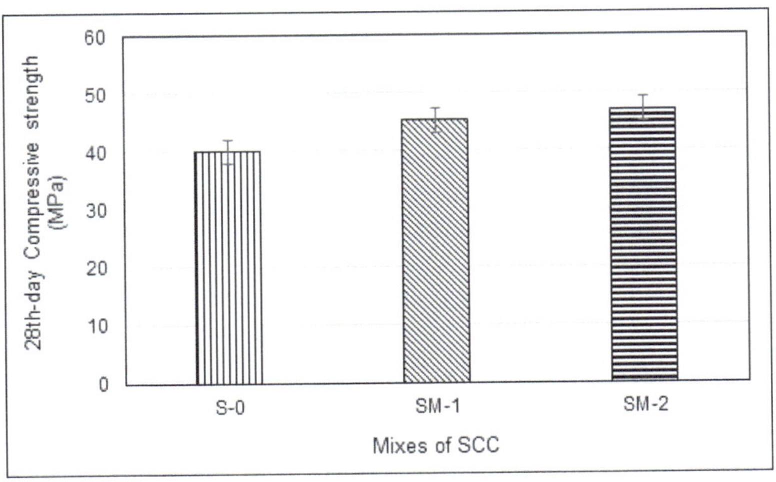

Figure 2.1 Compressive strength of conventional SCC and the SCC mixes containing mine tailing powder and nanocarbon dots.

The compressive strengths of all mixes containing mine tailings and nanocarbon dots tend to increase. The greatest compressive strength of a mix containing mine tailings and nanocarbon dots was at 15% replacement of OPC with mine tailing powder. This increase in compressive strength may be due to the higher fineness of mine tailing powder and the nano size of carbon dots, which affects the strength activator for cement-based materials. Moreover, the silica oxide (SiO_2) in mine tailing powder reacts with free lime and carbon produced during the hydration process of cement to form calcium-silicate-hydrate (C-S-H) and calcium-silicate-carbon-hydrate (C-S-C-H) compounds, which improve compressive strength in cost-effective SCC [11, 12]. Of the mine tailing mixes, the mix with 15% mine tailing powder and nanocarbon dots gives the highest compressive strength (Fig. 2.1). The results suggest that high-strength self-compacting concrete can be produced by incorporating highly fine mine tailing powder and nanocarbon dots with an appropriate quantity of replacement in OPC. The use of mine tailing powder and nanocarbon dots for producing high-strength SCC is cheaper than producing conventional SCC. In addition, the use of SCMs as a partial replacement of OPC in SCC can reduce the quantity of OPC and VMA to make SCC favorable and much workable. As a result, cheap, sustainable, and environmentally friendly SCC can be produced to realize the SDGs [12].

2.3.3 Splitting Tensile Strength

The splitting strength is an indirect method of determining the tensile strength of concrete by placing a concrete cylinder at the horizontal axis, between the platens of a testing machine. Figure 2.2 shows the splitting tensile strengths of various SCC mixes at 28 days. The cylindrical specimens (sized 150 mm × 300 mm) were casted to evaluate the splitting tensile strength according to ASTM C496/C496M-17 [9] by using a universal testing machine. The control mixture of SCC has a splitting tensile strength of 3.7 MPa at 28 days. The splitting tensile strengths of all mixes containing mine tailings and nanocarbon dots tend to increase. The greatest splitting tensile strength of a mix containing mine tailings and nanocarbon dots was at a 15% replacement of OPC with mine tailing powder.

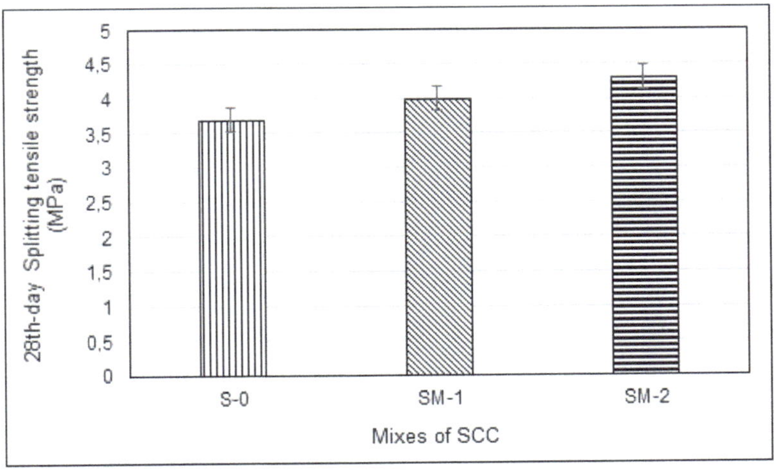

Figure 2.2 Splitting tensile strengths of various SCC mixes at 28 days.

This increase in splitting tensile strength may be due to the higher fineness of mine tailing powder and the nano size of carbon dots, which affects the strength activator for cement-based materials (Fig. 2.2). It has been observed that the splitting tensile strength of a self-compacting concrete mix containing mine tailing powder and nanocarbon dots gives higher values than the control SCC mix. The increase in splitting tensile strength is due to the pore refinement of self-compacting concrete resulting from the high fineness and pozzolanic action of SCMs and nanocarbon dots. The mix SM-3 has the highest splitting tensile strength among all the mixes (Fig. 2.2).

2.3.4 Flexural Strength

The flexural strengths of various SCC mixes at 28 days are shown in Fig. 2.3. Mortar, which is a material between cement paste and aggregates, has a highly complicated structure that not only answers to applied stresses but also depends on the interaction between its components. The inner structure of the mortar composite containing a wide range of aggregate stance consists of a bulk binder particle and supplementary particles, as this makes the morphology of the interfacial zone of transition (ITZ) much denser [13]. The ITZ, in turn, negatively affects the hardness, flexural strength, and surface capillarity in a binder-based composite with fine mortar sand if it has a lower density when compared to a conventional binder paste [13].

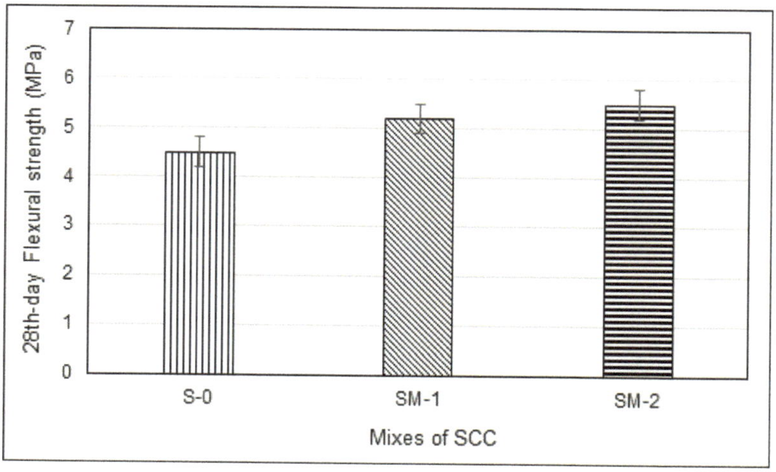

Figure 2.3 Flexural strengths of various SCC mixes.

It has been observed that the flexural strength of a self-compacting concrete mix containing mine tailing powder, granite waste powder, and corncob ash gives higher values than the control SCC mix. The flexural strength of the mix SM-3 showed the highest average (Fig. 2.3). To better understand the effect of nanocarbon dots on the flexural strength, a comparison between the flexural strengths of the mixes is presented here. The effect of adding nanocarbon dots on flexural strength was measured as a 4.5% increment for the 100% type 1 cement composite (SM-0), a 15.5% increment for the 95% type 1 cement composite (SM-1), and a 22.2% increase for the 85% type 1 cement composite (SM-3). These increments are in line with other research in which nanocarbon dots caused a higher flexural strength in zirconium [12], ultra-high performance concrete with fibers [14], self-consolidating cementitious systems [15, 16], roller concrete with waste aggregates [17], geopolymers [18], electrically conductive cementitious composites, concrete incorporating an iron particle–contained nanocarbon dot, and other construction binder–based materials [19–21]. These results can be attributed to the fact that a nanocarbon dot is tailored powder, which also contains carbon and could enhance the resiliency of the solid.

2.4 Conclusion

This study was conducted to produce sustainable and cost-effective SCC by using indigenous waste materials, where there is a huge potential for mining gemstones, but there is no availability of other SCMs such as fly ash, silica fume, and ground granulated blast furnace slag. Conclusions and recommendations drawn from the study are as follows:

- Mine tailing powder can be successfully used partially in place of OPC to make SCC cost effective and favorable.
- Since mine tailings are mechanically activated by grinding, grinding is beneficial for particle size reduction and homogenization, which enhances the pozzolanic action of mine tailings.
- Mine tailing powder, along with nanocarbon dots, increased the compressive strength by 17.7%, splitting tensile strength by 16.2%, and flexural strength by 22.2% of the control SCC mix at a replacement level of 15%.
- Successful use of mine tailings from quartz mines and nanocarbon dots has overcome the issue of environmental hazards due to the improper disposal of waste materials. Moreover, it would help in conserving natural resources and decrease the consumption of OPC in the making of SCC. In addition, it can make the construction industry viable and sustainable in terms of the UN SDGs.

Acknowledgment

The authors are thankful to the Department of Civil Engineering at Taxila and Karakorum International University, Pakistan, for providing their laboratory facilities to conduct this research study.

Author Contributions

MSK: Supervision, project administration, conceptualization, writing the original draft, methodology, writing review and editing; AGDSG: co-supervision, writing the original draft, methodology, writing review and editing; AE: co-supervision, methodology, software, writing review and editing; AA: data curation, visualization, resource, conceptualization, writing review and editing. All authors have read and agreed with the last stance of the manuscript.

Conflict of Interest

The authors declare that the research was conducted in the absence of any commercial or financial relationships that could be construed as a potential conflict of interest.

Data Availability

The data that support the findings of this study are available on request from the corresponding author [MSK].

Funding

This research did not receive any specific grant from funding agencies in the public, commercial, or not-for-profit sectors.

References

1. Kirgiz, M. S. Use of ultrafine marble and brick particles as alternative raw materials for clinkerization. *ZKG International* (2014), **4**, 36–44.
2. Kirgiz, M. S. Green cement composite concept reinforced by graphite nano-engineered particle suspension for infrastructure renewal material. *Composites Part B* (2018), **154**(12), 423–429.
3. Kirgiz, M. S. Advance treatment by nanographite for pulverized fly ash cement (the class F) systems. *Composites Part* B (2015), **82**(12), 59–71.
4. ASTM International. ASTM standard C150-07. Standard Specification for Portland Cement. 2007.
5. Kirgiz, M. S. Advancements in properties of cement containing pulverised fly ash and nanomaterials by blending and ultrasonication method (Review-Part I). *Nano Hybrids and Composites* (2018), **19**, 1–11.
6. Kirgiz, M. S. Advancements in properties of cement containing pulverised fly ash and nanomaterials by blending and ultrasonication method (Review-Part II). *Nano Hybrids and Composites* (2019), **24**, 37–44.

7. Turkish Standards Institution (TSE). *Methods of Testing Cement—Part 2: Chemical Analysis of Cement.* Ankara, Turkey (2002), pp. 1–27.
8. ASTM International. ASTM C39/C39M-21. *Standard Test Method for Compressive Strength of Cylindrical Concrete Specimens.* 2021.
9. ASTM International. ASTM C496/C496M-17. *Standard Test Method for Splitting Tensile Strength of Cylindrical Concrete Specimens.* 2017.
10. ASTM International. ASTM C618-19. *Standard Specification for Coal Fly Ash and Raw or Calcined Natural Pozzolan for use in Concrete.* 2019.
11. Kirgiz, M. S. Advancements in mechanical and physical properties for marble powder–cement composites strengthened by nanostructured graphite particles. *Mechanics of Materials* (2016), **92**, 223–234.
12. Kirgiz, M. S., Mirza, J., Cuc, S., Prodan, D., Saroşi, C., Perhaiţă, I., *et al.* Physico-antibacterial feature and SEM morphology of bio-hydraulic lime mortars incorporating nano-graphene oxide and binary combination of nano-graphene oxide with nano silver, fly ash, zinc, and titanium powders. *Buildings* (2023), **13**(1), 172.
13. Kirgiz, M. S. Nano size particle packing for smart nanoconcretes and cement-based materials: Theory and technology. In *Smart Nanoconcretes and Cement-Based Materials*, M. S. Liew, P. Nguyen-Tri, T. A. Nguyen, and S. Kakooei (Eds.). Elsevier, pp. 41–76, 2020.
14. Katman, H. Y. B., Khai, W. J., Bheel, N., Kirgiz, M. S., Kumar, A., and Benjeddou, O. Fabrication and characterization of cement-based hybrid concrete containing coir fiber for advancing concrete construction. *Buildings* (2022), **12**, 1450.
15. Hammat, S., Menadi, B., Kenai, S., Kirgiz, M. S., Carlos, T., Galdino, A. G. S., and Khatib, J. The effect of content and fineness of natural pozzolana on the rheological, mechanical, and durability properties of self-compacting mortar. *Journal of Building Engineering* (2021), **44**, 103276.
16. Katman, H. Y. B., Khai, W. J., Bheel, N., Kirgiz, M. S., Kumar, A., Khatib, J., and Benjeddou, O. Workability, strength, modulus of elasticity, and permeability feature of wheat straw ash-incorporated hydraulic cement concrete. *Buildings* (2022), **12**, 1363.
17. Ashteyat, A. M., Al Rjoub, Y. S., Obaidat, A. T., Kirgiz, M. S., Abdel-Jaber, M., and Smadi, A. Roller compacted concrete with oil shale ash as a replacement of cement: Mechanical and durability behavior. *International Journal of Pavement Research and Technology* (2022), **17**, 151–168.

18. Biricik, H., Kirgiz, M. S., de Galdino, A. G. S., Kenai, S., Mirza, J., Kinuthia, J., *et al.* Activation of slag through a combination of NaOH/NaS alkali for transforming it into geopolymer slag binder mortar: Assessment the effects of two different Blaine fines and three different curing conditions. *Journal of Materials Research and Technology* (2021), **14**, 1569–1584.

19. Kirgiz, M. S. Pulverized fuel ash cement activated by nanographite. *ACI Materials* (2018), **115**(6), 803–812.

20. Kirgiz, M. S. Effect of mineralogical substitution raw material mixing ratio on mechanical properties of concrete. *ZKG International* (2018), **10**, 30–41.

21. Sridhar, J., Gobinath, R., and Kirgiz, M. S. Comparative study for efficacy of chemically treated jute fiber and bamboo fiber on the properties of reinforced concrete beams. *Journal of Natural Fibers* (2022), **19**(15), 12224–12234.

Chapter 3

Valorization of Nanocarbon Dot–Based Green Binder Systems in Terms of Earthquake Engineering, Reinforced Concrete Building, Applied Mechanics, Transportation Engineering, and Their Applications

Mehmet Serkan Kirgiz,[a] **André Gustavo de Sousa Galdino,**[b] **Raul D. S. G. Campilho,**[c] **Rajesh Kumar,**[d] **Konstantinos G. Kolovos,**[e] **and Vinod M. Kumar**[f]

[a]*Academia of ACMCEN, Beaverton, OR 97075, USA*
[b]*Federal Institute of Education, Sciences and Technology of Espírito Santo, Av. Vitória, 1729, Jucutuquara, Vitória, ES, 29040-780, Brazil*
[c]*INEGI – Pólo FEUP, Rua Dr. Roberto Frias, 400, 4200-465 Porto, Portugal*
[d]*CSIR-Central Building Research Institute, Roorkee, Uttarakhand, 247667, India*
[e]*Department of Military Sciences, Division of Physical Sciences and Applications, Hellenic Army Academy, Vari, Greece*
[f]*Department of Civil Engineering, Vel Tech Rangarajan Dr. Sagunthala R&D Institute of Science and Technology, Chennai, India*
nakres42@yahoo.com

Handbook of Innovative Adhesive Technology: Biodegradable and Cost-Effective Binder for Built Environment and Engineering Sciences
Edited by Mehmet Serkan Kirgiz

ISBN 978-981-5129-51-9 (Hardcover), 978-1-003-59892-3 (eBook)
www.jennystanford.com

The aim of this chapter is comparing the nanotechnology-based green cement systems for earthquake engineering and reinforced concrete engineering. The experimented properties of the systems are water absorption, compressive strength, and flexural strength. Additionally, the mechanisms underlying the influence of nanocarbon dots (NCDs) on the properties of systems are discussed in detail in the chapter. Results reveal that NCDs allow effective combination of ground marble particle, class C fuel ash, and class F fuel ash with hydraulic Portland cement. Therefore, as summarized in this chapter, the process can be utilized in the large-scale manufacturing of green hydraulic cement along with NCDs for the current applications of earthquake engineering and reinforced concrete engineering.

3.1 Introduction

Marble powder (MP), class C fuel ash (CFA), and class F fuel ash (FFA) are the most known cementitious by-products generated by various industries such as quarry and energy. As the by-products contain precious chemical compounds, there is a need to use them in view of their activation effects. However, ordinary cement makes a big portlandite mass in the solid structure of mortar, stucco, and concrete. This big mass should be activated by supplementary materials, such as NCDs, class C fuel ash, class F fuel ash, and marble powder, to have better physical and mechanical properties in mortar, stucco, and concrete [1]. Recent advancement has significantly suggested an effective activation method for cement-based materials by nanoparticles (NPs) such as nano TiO_2 (nT), nano ZrO_2 (nZ), nano $CaCO_3$ (nC), nano limestone (nL), carbon nanotube (CNT), nano clay (nCL), and nano SiO_2 (nS) since 1999 [2–7]. Nanoparticles provide some benefits such as activation of cement hydration, which can reduce portlandite ($Ca(OH)_2$) formation and boost strength gain [8–14]. As the addition of nanoparticles does not provide an effective increase for the reduced early strength of green binder systems, the study focuses on NCDs to compare physical and mechanical properties of each other in green cement systems for earthquake engineering and reinforced concrete engineering.

3.1.1 Aim of This Study

This innovative research compares the results of physical tests and strength-related experiments on new cement and mortar. The effect of NCDs, marble powder, class C fuel ash, and class F fuel ash on water absorption, compressive strength, and flexural moment strength of mortar is expressed in view of the changes in the ratio of the constituent materials used for mortar preparation. It does not address either the properties of concrete made with marble powder, class C fuel ash, class F fuel ash, and NCDs or the substantial environmental effect of using by-products in composite materials. The partial replacement of cement by marble powder and class C fuel ash and class F fuel ash reduces the volume of wastes, but the main concern of this chapter is the need for comparing the influence of NCDs, marble powder, class C fuel ash, and class F fuel ash on water absorption, compressive strength, and flexural strength of composite materials. Therefore, the original research aims at monitoring water absorption and gain in the fundamental strength of mortar composites containing NCDs, marble powder, class C fuel ash, and class F fuel ash. These can be referred to enhance the performance of earthquake engineering and reinforced concrete engineering and to reduce the quantity of wastes and pure Portland cement, which ensures economic sustainability for construction and infrastructure applications.

3.2 Experimental Procedure

3.2.1 Preparation of Green Cement and Green Composite Materials for Testing

Table 3.1 presents the types of new binder, titles of sample, mixture, and blending ratio of cement constituent.

3.2.2 Determination of Water Absorption

Standard conditions required to measure the water absorption of mortar include 20 ± 3°C water temperature, 90 ± 5% relative humidity, and less than 0.2 m/s air speed in the laboratory [15].

The experiment unit is conditioned for 7 days in these standard conditions. In the experiment of water absorption, the following procedure uses a humidity-controlled curing cabinet for 7 days: (1) Mix the mortar according to Table 3.1; (2) after 1-day casting of the mortar, immerse the hardened mortar into water in the humidity-controlled curing cabinet; (3) add water to maintain the water level constant when necessary; (4) pull out the mortar from the water-curing cabinet on the seventh day; (5) place the mortar in a kiln on a tray at 50 ± 5°C; (6) desiccate the mortar until a constant weight is achieved; (7) monitor the water absorption of mortar by weighing in a scale at a precision of 0.01 in grams; (8) examine three samples for each mortar. Equation (3.1) computes the water absorption of mortar, and Table 3.2 presents the average of nine samples as descriptive water absorption.

Table 3.1 Types and codes of new cement, mixture, and blending ratio of cement constituent

Types of New Binder Material	Titles of Material	Mixture and Blending Ratio of New Cement Constituent (%)					
		CEM I 42.5N	Gypsum	MP	CFA	FFA	Carbon nano dot
Ordinary Portland cement (OPC)	[1]	100	0	0	0	0	0
MP blended-OPC	[2]	60	5	35	0	0	0
MP substituted-OPC	[3]	65	0	35	0	0	0
MP substituted-OPC-NCD	MP–C1	65	0	35	0	0	0.22
CFA substituted-OPC	CFA-C1	65	0	0	35	0	0
CFA substituted-OPC-NCD	CFA-C2	65	0	0	35	0	0.22
FFA substituted-OPC	FFA-C1	65	0	0	0	35	0
FFA substituted-OPC-NCD	FFA-C1	65	0	0	0	35	0.22

MP: marble powder; CFA: class C fuel ash; FFA: class F fuel ash; NCD: nanocarbon dot

$$W_{wa} = m_t - m_d \quad (3.1)$$

where W_{wa} is water absorption (in grams), m_t is the mass of the specimen after being dipped in water (in grams), and m_d is the mass of the dry specimen (in grams).

3.2.3 Determination of Compressive Strength

The TS EN 196-1 standard is employed to monitor the first-, third-, and seventh-day compressive strengths of the cubic mortar utilized in portions of the prism subjected to flexural moment experimentation [16]. A hydraulic compression machine tests three samples for each mix at each age, and Table 3.2 presents the average of the samples as the descriptive compressive strength.

Table 3.2 Types of material for earthquake engineering and reinforced concrete engineering, the average results of technological properties such as water absorption, compressive strength, and flexural strength at first, third, and seventh days

Types of Material	Technological Properties						
	Water Absorption (%)	Compressive Strength (MPa)			Flexural Strength (MPa)		
		Day 1	Day 3	Day 7	Day 1	Day 3	Day 7
OPC (Reference)	N/A	N/A	N/A	34.2	N/A	N/A	N/A
MP blended-OPC	N/A	N/A	N/A	21.8	N/A	N/A	4.1
MP substituted-OPC	N/A	N/A	N/A	22.7	N/A	N/A	N/A
MP substituted-OPC-NCD	2.3	N/A	N/A	37.6	N/A	N/A	5.2
CFA substituted-OPC	7	4.2	9.6	N/A	1.8	3.2	N/A
CFA substituted-OPC-NCD	9.2	3.7	9.1	N/A	2.5	4	N/A
FFA substituted-OPC	9.4	5.7	8	N/A	1.9	3.3	N/A
FFA substituted-OPC-NCD	9.5	4.9	10.1	N/A	2.4	3.8	N/A

N/A means that the measurement is not available to carry out; MP: marble powder; OPC: ordinary Portland cement; CFA: class C fuel ash; FFA: class F fuel ash; NCD: nanocarbon dot

3.2.4 Determination of Flexure Moment Strength

The TS EN 196-1 standard is used to monitor the first-, third-, and seventh-day flexural moment strengths of the prism mortar [16]. A hydraulic bending machine tests three samples for each mix at each age, and Table 3.2 presents the average of the samples as the descriptive flexural moment strength.

3.3 Results

Table 3.2 gives the types of mortar and the average results of the physical and mechanical properties of mortar, such as water absorption, compressive strength, and flexural strength at first, third, and seventh days.

3.4 Discussions

3.4.1 Water Absorption of Material

Table 3.2 presents the average of water absorption for mortar containing marble powder, class C fuel ash, and class F fuel ash (as seen in Fig. 3.1). Since the study increases the quantity of the ground by-product in cement and mortar, this growth causes a fluctuation in the water absorption of mortar. Results show that the content of NCD up to 0.22% overcomes this fluctuation in the mortar (Table 3.2). Therefore, it can be suggested that prolonged curing, increasing age, and percentage of NCD can lead to a reduction in the permeable voids of the mortar. This is due to the high activation effects of NCD on the physical properties of mortar. Figure 3.1 presents the changings of water absorption (as a percentage) in the tested mortar.

Although this study reduces the amount of pure Portland cement (by up to 35%) in mortar, the water absorption decreases since the mortar contains 0.22% NCD (Table 3.2 and Fig. 3.1). The results also reveal the fact that as mortar includes 0.22% NCD, water absorption decreases. In comparison with the results, one notes that the by-product in the mortar system needs 0.22% NCD for properly combining the by-product and cement grains. Another result could be that the interfacial transition zone in mortar is enhanced because of high reactivity as well as the activation effect of NCD [1, 2]. It is shown that "by addition of nanoparticle, the total specific pore volume of cement based material decreased, and the most probable pore diameter shifts to smaller pores and falls in the range of few harm pore," which indicates that the addition of nanoparticles fills the pore structure and, hence, the water absorption is reduced [1, 2]. The addition of NCD and by-product content may form some flocculation due to the replacement of cement with the by-product. This weakens the apparent pore structures of green mortar.

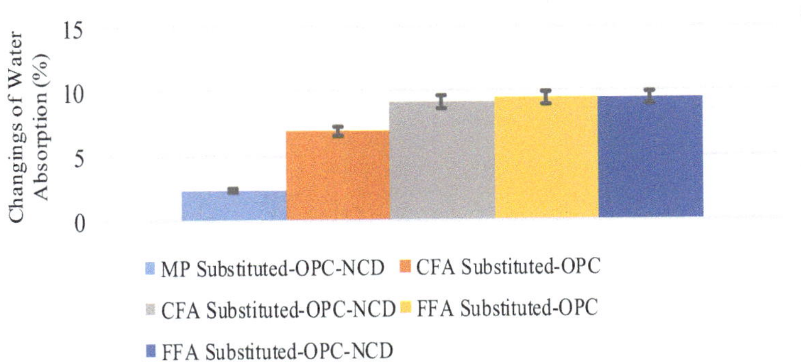

Figure 3.1 Changings of water absorption (as a percentage) in various mortar samples containing 35% by-product, such as marble powder, class C fuel ash, and class F fuel ash.

3.4.2 Changings of Compressive Strength Gain of Material

The regular axial compression force applied to the outer field of the mortar sample monitors the compressive strength of mortar containing marble powder, class C fuel ash, class F fuel ash, NCD, and pure Portland cement at 1d, 3d, and 7d. Figure 3.2 presents the changings of compressive strength at 1d, 3d, and 7d, strength percentage bar, the substituted cement mortar, and pure Portland cement mortar.

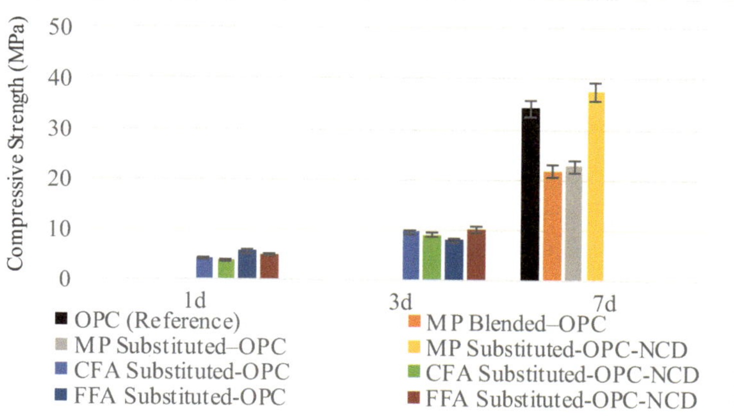

Figure 3.2 Changings of compressive strength at 1d, 3d, and 7d, strength percentage bar, the substituted cement mortar, and pure Portland cement mortar.

The water-to-binder ratio is stable at 0.5 for preventing the positive effect of the reduced W/B on strength gain. However, closed curing cabinet is not done by warm water so that the strength of mortar is not positively influenced (Tables 3.1 and 3.2 and Fig. 3.2). The effort monitors the gain in standard compressive strength for a green cement–mortar system with the addition of NCD.

3.4.3 Changings of Flexural Moment Strength Gain of Material

The regular axial flexure force applied to the outer field of the mortar sample monitors the flexural moment strength of mortar containing marble powder, class C fuel ash, class F fuel ash, NCD, and pure Portland cement at 1d, 3d, and 7d. Figure 3.3 presents the changings of flexural moment strength at 1d, 3d, and 7d, substituted mortar, and pure Portland cement mortar.

The water-to-binder ratio is stable at 0.5 for preventing the positive effect of the reduced W/B on the flexural moment strength gain. However, closed curing cabinet is not done by warm water so that the flexural moment strength of mortar is not positively influenced (Tables 3.1 and 3.2, and Fig. 3.3). Consequently, the flexural moment strength of the green cement–mortar system was enhanced by the incorporation of NCD.

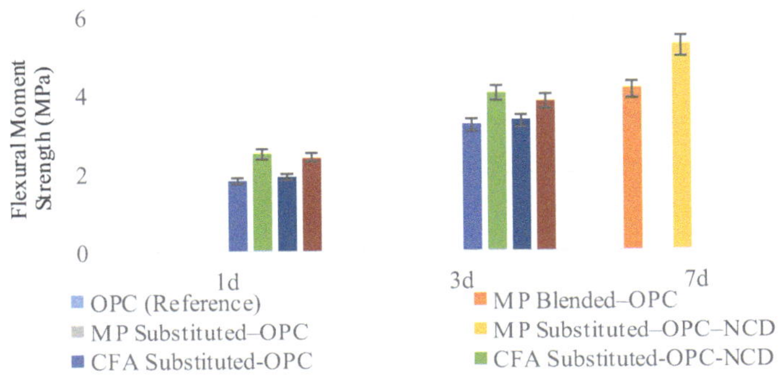

Figure 3.3 Changings of the flexural moment strength at 1d, 3d, and 7d, strength percentage bar, the substituted cement mortar, and pure Portland cement mortar.

3.5 Conclusion

The authors carried out this laboratory study to present more effective insights into the effects of NCDs on the physical and mechanical properties of green cement systems. This presentation also infers conclusion from the current experimental study to present them properly. The results support the following insight:

The results of the experimental investigation indicate that NCDs enhance the compressive strength gain and the flexural moment strength gain in green cement systems, including marble powder, class C fuel ash, and class F fuel ash, in a successful manner. Furthermore, the results demonstrate that NCDs also reduce the water absorption of these systems. Additionally, NCDs can also be described as an increaser of strength gain and a reducer of water absorption for hydraulic green cement systems. In view of the earlier conclusions, earthquake engineering and reinforced concrete engineering can use NCDs as prepared in this non-budget research method for the large-scale production of hydraulic green cement containing marble powder, class C fuel ash, and class F fuel ash.

References

1. Kirgiz, M. S. (2007). *Usage of the Wastes of Marble and Brick Industries in Cement Manufacturing as Mineralogical Additive.* Philosophy of Doctorate Thesis, University of Gazi, Graduate School of Natural and Applied Sciences, Ankara. (https://tez.yok.gov.tr/UlusalTezMerkezi).
2. Kirgiz, M. S. (2016). Advancements in mechanical and physical properties for marble powder–cement composites strengthened by nanostructured graphite particles. *Mechanics of Materials,* **92**(1), 223–234. DOI: 10.1016/j.mechmat.2015.09.013.
3. Kirgiz, M. S. (2015). Advance treatment by nanographite for Portland pulverised fly ash cement (the class F) systems. *Composites Part B,* **82**(12), 59–71. DOI: 10.1016/j.compositesb.2015.08.003.
4. Kirgiz, M. S. (2015). Supernatant nanographite solution for advance treatment of C class fly ash–cement systems (Part 2). *ZKG International,* **5**, 42–47.
5. Kirgiz, M. S. (2015). Supernatant nanographite solution for advance treatment of C class fly ash–cement systems (Part 1). *ZKG International,* **4**, 56–65.
6. Kirgiz, M. S. (2015). Advances in physical properties of C class fly ash–cement systems blended nanographite (Part 2). *ZKG International,* **1–2**, 60–67.
7. Kirgiz, M. S. (2014). Advances in physical properties of C class fly ash–cement systems blended nanographite (Part 1). *ZKG International,* **12**, 42–48.
8. Kirgiz, M. S. (2017). Experimental study on pH-assisted bending strength gain projection for green mortar composite containing brick powder as substitution by-product. *European Journal of Environmental and Civil Engineering,* Published Online, DOI: 10.1080/19648189.2017.1357740.
9. Kirgiz, M. S. (2011). Chemical properties of blended cement pastes. *Journal of Construction Engineering and Management,* **137**, 1036–1042.
10. Kirgiz, M. S. (2014). Effects of blended-cement paste chemical composition changes on some strength gains of blended-mortars. *The Scientific World Journal,* **1**, 1–8.
11. Kirgiz, M. S. (2015). Strength gain mechanisms of blended-cements containing marble powder and brick powder. *KSCE Journal of Civil Engineering,* **19**(1), 165–172.

12. Kirgiz, M. S. (2016). Strength gain mechanism for green mortar substituted marble powder and brick powder for Portland cement. *European Journal of Environmental and Civil Engineering*, **20** (sup1), 38–63. DOI: 10.1080/19648189.2016.1246691.
13. Kirgiz, M. S. (2016). Fresh and hardened properties of green binder concrete containing marble powder and brick powder. *European Journal of Environmental and Civil Engineering*, **20** (sup1), 64–101. DOI: 10.1080/19648189.2016.1246692.
14. Kirgiz, M. S. (2015). Use of ultrafine marble and brick particles as raw materials in cement manufacturing. *Materials and Structures*, **48**(9), 2929–2941. DOI: 10.1617/s11527-014-0368-6.
15. Turkish Standards Institute (TSI). (2016). *TS 699/T1: Natural Buildings Stones-Methods of Inspection and Laboratory Testing*, TSI Publication, Turkey, pp. 1–42.
16. Turkish Standards Institute (TSI). (2016). *TS EN 196-1: Methods of Testing Cement- Part 1: Determination of Strength*, TSI Publication, Turkey, pp. 1–35.

Chapter 4

Some Technological Properties of Hydraulic Construction Lime and Grouting Incorporated with Chopped Carbon Nanotube Fibers for Manufacturing Cost-Effective and CO_2-Neutral Materials in Cleaner Built Environment

Mehmet Serkan Kirgiz,[a] Andre G. D. S. Galdino,[b] John Kinuthia,[c] Raul D. S. G. Campilho,[d] Roberto A. Gonzalez-Lezcano,[e] and Konstantinos G. Kolovos[f]

[a]*Academia of ACMCEN, Beaverton, OR 97075, USA*
[b]*Federal Institute of Education, Sciences and Technology of Espírito Santo, Av. Vitória, 1729, Jucutuquara, Vitória, ES, 29040-780, Brazil*
[c]*Advanced Materials Testing Centre (AMTeC), School of Engineering, University of South Wales, United Kingdom*
[d]*INEGI – Pólo FEUP, Rua Dr. Roberto Frias, 400, 4200-465 Porto, Portugal*
[e]*Architecture and Design Department, Escuela Politécnica Superior, Universidad CEU San Pablo, CEU Universities, 28040 Madrid, Spain*
[f]*Department of Military Sciences, Division of Physical Sciences and Applications, Hellenic Army Academy, Vari, Greece*
nakres42@yahoo.com

Handbook of Innovative Adhesive Technology: Biodegradable and Cost-Effective Binder for Built Environment and Engineering Sciences
Edited by Mehmet Serkan Kirgiz

ISBN 978-981-5129-51-9 (Hardcover), 978-1-003-59892-3 (eBook)
www.jennystanford.com

The use of chopped carbon nanotube fibers (CCNTFs) as a supplement in the making of hydraulic lime bed grouting is of great interest to the scientific and construction technology community because of their activation effect on the properties of building hydraulic lime. The effort monitored the impact of CCNTFs on some properties of classic hydraulic lime grouting using a mixture ratio of 1:3 CL90S lime and fine sand and mixing water. The mixing proportion of control grouting and new CCNTF grouting was kept constant. The tested properties were chemical components of CL90S lime, some physical properties of CL90S, as well as apparent porosity and water hydrophobicity of new CCNTF grouting. The best results were determined for the grout with 5% CCNTF in terms of apparent porosity and water hydrophobicity. And this could improve the experimental properties of grouting, which can be used for construction technology.

4.1 Introduction

Hydraulic lime has been a highly utilized material in both new and existing buildings for centuries. Its applications include masonry grouting, maintenance of monuments, and the preservation of heritage buildings [1]. Lime grout is able to breathe and allow the passing of air and moisture in and out. It also well absorbs carbon dioxide (CO_2) from air. It is also noteworthy that natural hydraulic lime can react with water and gain initial strength before undergoing a period of prolonged development, during which it absorbs CO_2. These features have made hydraulic lime favorable since Portland cement was discovered [2]. Despite these unique features of hydraulic lime, there are some deficits that need to be overcome. For example, hydraulic lime absorbs too much liquid, has a lower specific surface area, and has greater porosity when compared to other binders in the construction industry. Because of these deficits, natural hydraulic lime had lost its popularity and was being used in unimportant production in the construction industry. To update its usage in a number of industries, comprehensive research was carried out. Kirgiz *et al.* has reported that graphite nano powder (GNP) has great impact on the strength and non-destructive properties of hydraulic lime grout [3]. In a separate study, Dorin *et al.* demonstrated the efficacy of graphene oxide nano powder (GONP) on the characteristics of hydraulic lime. They demonstrated that the incorporation of GONP resulted in a significant enhancement in the adhesion of grout to the brick substrate [4]. In another work, Kirgiz *et al.* indicated that the existence of GNP accelerated the gain of both compression and bending strengths in binder-based materials [5]. After Kirgiz *et al.* and Dorin *et al.*, many construction materials used GNP and GONP as additives in making self-consolidating binder materials [6], composites with coarse aggregates of e-waste plastic [7], geo-polymer [8], electrically conductive cementitious composites [9], the composite which was formed by incorporating the nanographite by-product [10], reinforcement of binders [11], and high-volume fly ash–blended concrete including graphene nanoplatelets [12]. Moreover, one of the aforementioned efforts [6] reported that the addition of 0.05% GNP by mass of binder resulted in 57% and 48% increase in both compressive and flexural strengths, respectively. In light of the studies, the aim of this research is to develop hydrophobic, porosity, and specific surface characteristics of hydraulic lime grout with various percentages of GONP.

The manuscript presents a novel approach to hydraulic grouting, in which CCNTFs are incorporated. This approach was evaluated in terms of porosity, Blaine surface, water absorption, and coefficient equations between the properties of bed grouting.

4.2 Materials and Methods

4.2.1 Materials in Making Grout Sample

Hydraulic lime, sand, water, and supplements, such as CCNTFs, are the preparation materials of bed grouting for this research. Natural hydraulic lime, also known as CL90S, was used, which was made by Nuh Construction Products Co., complied with specifications in the TS EN 459-1 standard [13]. A standard fine grouting sand was used along with particle size under 2 mm. CCNTFs have a size range of approximately twenty to ten microns.

4.2.2 Preparation of Grouting

For the preparation of grouting with CCNTFs and without CCNTFs (known as the control sample), the binder-to-aggregate ratio and the water-to-binder ratio were kept constant at 1:3 and 0.5, respectively.

In the control group, the grouting sample includes no CCNTFs. Three other types (CL90S grouting) were prepared containing 1%, 2.5%, and 5% CCNTFs by the weight of lime. Table 4.1 shows the types of grouting, groups of grouting, mixture composition of the control grouting, and new CCNTF-CL90S grouting and its mixture characteristics.

Table 4.1 Types of grouting, groups of grouting, mixture composition of the control grouting, and new CCNTF-CL90S grouting and its mixture characteristics

Groups	Types of Grouting	Lime to Sand (v/v)	Water to Lime (wt. %)	CCNTF to lime (wt. %)
Control grouting	CL90S	1/3	0.5	0
New grouting	CCNTF-CL90S 1%	1/3	0.5	0.01
	CCNTF-CL90S 2.5%	1/3	0.5	0.025
	CCNTF-CL90S 5%	1/3	0.5	0.05

The control grouting sample CL90S and the group of samples with CCNTFs were prepared according to the following steps. The quantities were previously weighed in the laboratory scale. An effective mixing of the three components—lime, sand, and CCNTFs—was carried out in a stirrer for 15 min. The grout was made in the same step as in the case of any grout, as mentioned in Table 4.2, and blending of water gradually.

4.2.3 Testing

According to the ASTM C25-19 [14] and ASTMC110-20 [15] standards, the chemical components and physical specifications of CL90S were determined. Moreover, water retention and apparent porosity tests were, respectively, performed as per ASTM C110-20 [15] and TS 699/T1 [16].

4.3 Results and Discussion

4.3.1 Chemical Components of CL90S

Table 4.2 shows the chemical components of CL90S, as well as the standard specifications that CL90S must have according to ASTMC

911–19 [17]. The various chemical compounds that are collectively referred to as "lime," including quicklime, hydrated lime (slacked lime), and limestone, are primarily produced through the heating of limestone in a kiln. Hydrated lime is manufactured by combining quicklime with water to transform the CaO powder into a slurry, which is then converted into a flowable calcium hydroxide ($Ca(OH)_2$).

Table 4.2 Chemical components of CL90S, as well as standard specifications that CL90S must have according to ASTMC 911–06

Chemical Components	**CL90S**	**Standard Specification as ASTMC 911-06**
Calcium hydroxide ($Ca(OH)_2$) (%)	90	—
Calcium oxide (CaO) + Magnesium oxide (MgO) (%)	91	>90
Magnesium oxide (MgO) (%)	3	≤5
Sulfate oxide (SO_3) (%)	2	≤2
Loss on ignition (%)	4	≤4
Total	100	100

From being used in the medieval ages, there are a plenty of usage areas for lime compounds today, e.g., in iron and steel manufacturing, building construction, wastewater treatment, pulp and paper production, and agriculture. Calcium hydroxide has a polymeric structure, as do all metal hydroxides. The structure is identical to that of $Mg(OH)_2$ (brucite structure), i.e., the cadmium iodide phenomena. A weak hydrogen bond exists between its layers [18]. In the laboratory, it can be prepared by mixing aqueous solutions of calcium chloride and sodium hydroxide (as in Table 4.2). The mineral form, portlandite, is relatively rare but can be found in some volcanic, plutonic, and metamorphic rocks. It has also been known to arise in burning coal dumps. As could be understood from Table 4.1, the lime used (known as CL90S) has no portlandite compound in its inner structure. The positively charged ionized species $CaOH^+$ has been detected in the atmosphere of S-type stars [19]. In other words, the earth also had calcium hydroxide in its atmosphere. Both the chemical properties of CL90S comply with the specifications in the ASTMC 911–06 standard.

4.3.2 Physical Properties of CL90S

Another standard—AS 1672-1997—explains specifications and testing for building lime [20]. The specifications set forth the principal physical requisites for high calcium hydrated lime. In terms of physical characteristics, high calcium hydrated lime must undergo sieve analysis. The available lime index (ALI) shows how much lime is available for reaction in building materials. This index is vital for users, as it shows how much of the lime reacted with the mixing water. Besides, the available lime is always less than the total lime. Total lime includes lime combined as carbonate, silicate, and other forms not available for chemical combination [20]. Considering the knowledge that AS 1672-1997 represented, the physical quality of lime is essential for preventing harmful chemicals in the environment, and it is also important for the processes by which limestone is broken down into quicklime. In addition, quicklime is slaked into hydrated lime [18, 20]. Table 4.3 shows the physical properties of CL90S, as well as the standard specifications that CL90S must have according to ASTMC 911–06.

Table 4.3 Physical properties of CL90S, as well as standard specifications that CL90S must have according to ASTMC 911–06

Physical Properties		CL90S	Standard Specification as per ASTMC 911–06
Moisture (%)		1	<2
Bulk density (kg/m^3)		500	≤600
Residue on sieve (%)	0.200 (mm)	1	<2
	0.090 (mm)	5	<7
Blaine fineness (cm^2/g)		650	—

The residue obtained from the sieve test is prepared in the following manner. A 100-gram sample of hydrated lime is subjected to a sieve test. The test includes sieves arranged in such a way that the coarser sieves are placed at the top in a decreasing order. A moderate shaking system facilitates the transfer of hydrated lime through the sieves from the top in a period of no more than 30 minutes. The residue on each of the sieves is weighed at 21 ± 2°C until a constant mass is achieved. The percentage mass of the hydrated lime sample is presented in Table 4.3. Both the physical properties of CL90S comply with the specifications of the ASTMC 911-06 standard.

4.3.3 Apparent Porosity

Figure 4.1 shows the results related to the apparent porosity of the control sample and the CCNTF sample. Considering the conventional cement-based grouting, hydraulic lime-based grouting varies importantly.

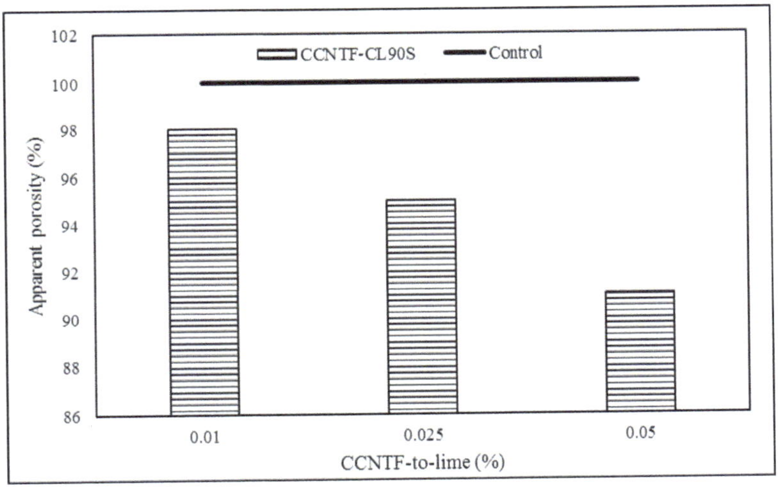

Figure 4.1 Results related to the apparent density of the control sample and the new CCNTF sample.

The binder matrix of new grouting, CCNTF-CL90S, is softer than the cement one, and the strength capacity is less than 10 MPa. The apparent specific density is less than 2 g/cm^3, while the cement's specific density is more than 3.1 g/cm^3. The hardening process of cement grouting could result in creating a net of micro- and macro-cracks. To measure the apparent porosity of the hydraulic lime-based grouting, which is a significant physical property that gives an insight into the level of strength and pathology stance of construction materials, different parameters should be considered [21]. Porosity takes out a great volume of this type of grouting. It is affected by many parameters such as the grout's technological point of view—type of binder used, the binder-to-sand ratio, the type and size of sand, the compaction applied during their placing, and the influence of the environment. Moreover, it does not have a rigid form, but it can be easily changed, majorly due to the action of time, and additionally the sizes of pores differ in a broad range, from nanometers to a few millimeters [22]. There are two main forms of porosity in the grouting structure: rounded with different sizes and elongated cracks [23]. In accordance with the International Union of Pure and Applied Chemistry (IUPAC) (1972) [24], pores are classified into three categories: micropores, mesopores, and macropores. Macropores exert a significant influence on strength, mesopores are associated with transfer properties, and micropores are responsible for absorption properties [25]. The dimensions of spherical pores can range from over 100 micrometers to several millimeters, and the entrapped air primarily contributes to the formation of spherical pores due to the absence of compaction during mixing. The aforementioned pores can be encountered in contact with an aggregate stack and within the binder phase. There are some cracks on the pore line letting in air and liquid. These pores with cracks seem to have a greater influence on mortar strength and durability [26]. From Fig. 4.1, it can be noted that CCNFTs reduce the capillary pores in the new CCNTF sample. Capillary pores of diameter smaller than 0.1 μm are formed in the binder matrix and in the interfacial transition zones [27] during the hydration and hardening process, which removes the surplus water. The water-to-lime ratio and the carbonation-to-hydration ratio also affect the percentage of capillary pores. The pores are reduced with time unless a secondary effect—drying-shrinkage, expansion, and chemical and mechanical processes due to the environment and air condition—occurs [28]. The apparent porosity of the tested materials decreases from 100% to 91% in the control sample and CCNTF sample. The incorporation of CCNTFs has a notable impact on the apparent porosity structure of the novel grouting system. However, the apparent density of the hardened grouting remains relatively unaltered. The lowest apparent porosity is represented by the CCNTF-CL90S 5% grouting with the presence of a CCNTF-to-lime ratio of 5% (Fig. 4.1). On the other hand, cracks can be the result of drying-shrinkage because of the differential movement of the grouting constituents during the setting time and hardening process of the lime-based binder materials. It is randomly oriented with the grouting inner structure, and its geometry differs in terms of the technological characteristics of the grouting and the pathology level.

4.3.4 Water Hydrophobicity

Figure 4.2 shows the results related to the water hydrophobicity of the control sample and the new CCNTF sample. With the supplement of CCNTF 5%, there was a major increase in water hydrophobicity. While in the case of a simple CCNTF supplement, for a concentration of 2.5%, the water absorption decreases than that of CCNTF-CL90S 1%. An increase in water hydrophobicity is a beneficial factor in preventing moisture and harmful gases from entering the wall and structural system. Mohammed *et al.* [29] demonstrated that grouting with 0.03% by weight of graphene exhibited a reduction in water absorption relative to the control grouting, due to the smaller pore quantity resulting from the addition of the graphene particle. Therefore, according to the results obtained by his research team, the capillary porosity was reduced.

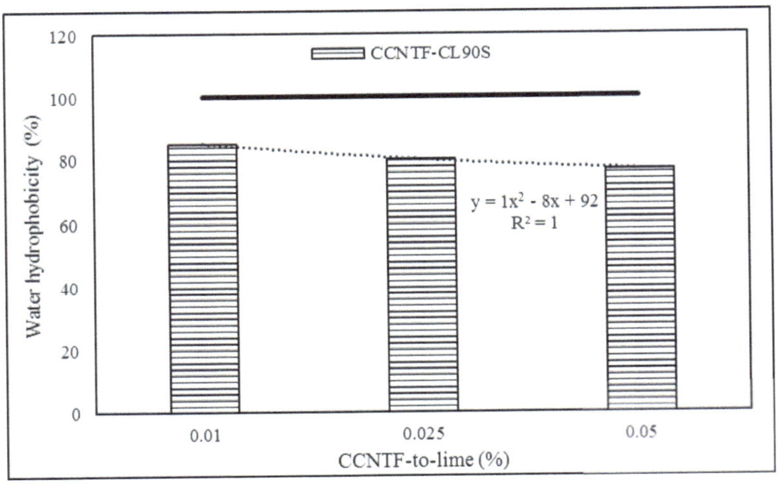

Figure 4.2 Results related to the water hydrophobicity of the control sample and the new CCNTF sample.

As illustrated in Fig. 4.2, the observed trend aligns with the findings of Mohammed et al. The researchers have proposed that the incorporation of graphene particles into cement composites can enhance their resistance to aggressive effects. This is achieved by strengthening the interfacial transition zone of the cement matrix with graphene particles, which serves as a robust barrier and reduces the penetration of harmful liquids and gases. Figure 4.2 also shows the regression formula and the R^2 coefficient between water hydrophobicity and the CCNTF-to-lime ratio. From Fig. 4.2, it can be seen that the highest value of the hydrophobicity coefficient was recorded as 1, which shows a strong relationship between the two properties.

4.4 Conclusion

New hydraulic lime grouting was manufactured by incorporating CCNTFs and was evaluated in terms of the chemical compound of lime (CL90S), physical property of lime, as well as apparent porosity and water hydrophobicity of the control grouting and the new grouting prepared with CCNTF 1%, 2.5%, and 5%. The results indicated that CCNTF-CL90S 5% grouting yielded more favorable outcomes, suggesting enhanced incorporation of CCNTFs with hydraulic lime. This resulted in a notable improvement in the apparent porosity structure and water absorption characteristics when compared to conventional lime grouting. A reduction in the weight percentage of CCNTF blended with the other components results in a slight decrease in both water absorption and apparent porosity. The grout provided effective results for hydrophobicity because absorption is slower and lower than that in classic hydraulic lime grouting.

Author Contribution

Conceptualization: MSK, RDSGC, JK, TT; methodology: MSK, AGDSG, RDSGC; writing—original draft: MSK, KGK; writing—review and editing: MSK, RDSGC, JK, OKD, KGK; supervision: MSK; visualization: JK, KGK; project administration: MSK, RDSGC; resources: OKD, TT, KGK; and data curation: MSK, TT, OKD, KGK.

Conflict of Interest

The authors declare that the research was conducted in the absence of any commercial or financial relationships that could be construed as a potential conflict of interest.

Data Availability

The data that support the findings of this study are available on request from the corresponding author (Mehmet Serkan Kirgiz).

References

1. Anonymous. Buxton Lime—Natural Hydraulic Lime, NHL, hydraulic lime, renders, plasters, mortars, restoration, refurbishment, air lime, building limes, limecrete, lime concrete, flooring, masonry, stone repair, building. Retrieved from: https://web.archive.org/web/20090322062702/; http://www.tarmac.co.uk/BuxtonLime/hydrauliclime.aspx. (08.06.2023).
2. Ashurst, J. (1997). The technology and use of hydraulic lime. *The Building Conservation Directory*, 1–5. Retrieved from: https://www.buildingconservation.com/articles/hydraulic/hydraulic.htm. (08.06.2023).
3. Katman, H. Y. B., Khai, W. J., Kirgiz, M. S., Nehdi, M. L., Benjeddou, O., Thomas, B. S., Papatzani, S., Rambhad, K., Kumbhalkar, M. A., and Karimipour, A. (2022). Transforming conventional construction binders and grouts into high-performance nanocarbon binders and grouts for today's constructions. *Buildings*, **12**, 1041. https://doi.org/10.3390/buildings12071041.
4. Kirgiz, M. S., Mirza, J., Cuc, S., Prodan, D., Saroşi, C., Perhaiţă, I., Carpa, R., Moldovan, M., Popa, D., Varvara, S., and Popa, S. (2023). Physico-antibacterial feature and SEM morphology of bio-hydraulic lime grout incorporating nano graphene oxide and binary combination of nano graphene oxide with nano silver, fly ash, zinc, and titanium powders. *Buildings*, **13**(1), 172. https://doi.org/10.3390/buildings13010172.
5. Kirgiz, M. S., de Sousa Galdino, A. G., Kinuthia, J., Khitab, A., Hassan, M. I. U., Khatib, J., El Naggar, H., Thomas, C., Mirza, J., Kenai, S., Nguyen, T. A., Nehdi, M., Syarif, M., Ashteyat, A., Gobinath, R., Soliman, A., Tagbor, T. A., Kumbhalkar, M. A., Bheel, N., and Tiwary, C. S. (2021). Synthesis, physico-mechanical properties, material processing, and math models of novel superior materials doped flake of carbon and colloid flake of carbon. *Journal of Materials Research and Technology*, **15**, 4993–5009.
6. Ullah, M., Burhan, S., Imtiazi, A., Khushnood, R. A., Erum Pervaiz, E., Ahmed, W., Ullah, A., and Qureshi, Z. A. (2021). Synthesis, characterization and application of graphene oxide in self-consolidating cementitious systems. *Construction and Building Materials*, **296**, 123623.

7. Ahmad, F., Qureshi, M. I., and Ahmad, Z. (2022). Influence of nano graphite platelets on the behavior of concrete with E-waste plastic coarse aggregates. *Construction and Building Materials*, **316**, 125980.
8. Jindal, B. B., and Sharma, R. (2020). The effect of nanomaterials on properties of geopolymers derived from industrial by-products: A state-of-the-art review. *Construction and Building Materials*, **252**, 119028.
9. Sun, J., Lin, S., Zhang, G., Sun, Y., Zhang, J., Chen, C., Morsy, A. M., and Wang, X. (2021). The effect of graphite and slag on electrical and mechanical properties of electrically conductive cementitious composites. *Construction and Building Materials*, **281**, 122606.
10. Dong, W., Huang, Y., Lehane, B., Aslania, F., and Ma, G. (2021). Mechanical and electrical properties of concrete incorporating an iron-particle contained nano-graphite by-product. *Construction and Building Materials*, **281**, 121377.
11. Sheikh, T. M., Anwar, M. P., Muthoosamy, K., Jaganathan, J., Chan, A., and Mohamed, A. A. (2021). The mechanics of carbon-based nanomaterials as cement reinforcement: A critical review. *Construction and Building Materials*, **303**, 124441.
12. Adamu, M., Trabanpruek, P., Jongvivatsakul, P., Likitlersuang, S., and Iwanami, M. (2021). Mechanical performance and optimization of high-volume fly ash concrete containing plastic wastes and graphene nanoplatelets using response surface methodology. *Construction and Building Materials*, **308**, 125085.
13. TSI. TS EN 459-1. (2015). *Building Lime—Part 1: Definitions, Specifications and Conformity Criteria*. Turkish Standard Institute.
14. ASTM. ASTM C25-19. (2019). *Standard Test Methods for Chemical Analysis of Limestone, Quicklime, and Hydrated Lime*. ASTM International.
15. ASTM. ASTM C110-20. (2020). *Standard Test Methods for Physical Testing of Quicklime, Hydrated Lime, and Limestone*. ASTM International.
16. TSI. TS 699/T1. (2016). *Natural Building Stones—Methods of Inspection and Laboratory Testing*. Turkish Standard Institute.
17. ASTM C911-19. (2019). *Standard Specification for Quicklime, Hydrated Lime, and Limestone for Selected Chemical and Industrial Uses*. ASTM International.
18. Greenwood, N. N., and Earnshaw, A. (1997). *Chemistry of the Elements* (2nd edn), Butterworth-Heinemann, Oxford. ISBN 0-7506-3365-4.

19. Jørgensen, U. G. (1997). Cool star models, in *Molecules in Astrophysics: Probes and Processes* (Ewine F. van Dishoeck, Ed.), Vol. 178, Springer Science & Business Media, p. 446, ISBN 079234538X.
20. ASI. AS 1672-1997. (1997). *Limes and Limestones: Limes for Building*. Australian Standard Institute.
21. Mallidi, S. R. (1996). Application of mercury intrusion porosimetry on clay bricks to assess freeze-thaw durability: A bibliography with abstracts. *Construction and Building Materials*, **10**(6), 461–465.
22. Diamond, S. (2000). Mercury porosimetry: An inappropriate method for the measurement of pore size distributions in cement-based materials. *Cement and Concrete Research*, **30**(10), 1517–1525.
23. Shi, D., and Winslow, D. N. (1985). Contact angle and damage during mercury intrusion into cement paste. *Cement and Concrete Research*, **15**(4), 645–654.
24. Anonymous. (2019). *Compendium of Chemical Terminology* (2nd edn, the "Gold Book"). Compiled by A. D. McNaught and A. Wilkinson, International Union of Pure and Applied Chemistry (IUPAC). Oxford: Blackwell Scientific Publications. Online version created by S. J. Chalk. ISBN 0-9678550-9-8. https://doi.org/10.1351/goldbook.
25. Elert, K., Rodriguez-Navarro, C., Pardo, E. S., Hansen, E., and Cazalla, O. (2002). Lime mortars for the conservation of historic buildings. *Studies in Conservation*, **47**(1), 62–75.
26. Maria, S. (2016). Use of natural pozzolans with lime for producing repair mortars. *Environmental Earth Sciences*, **75**, 758. https://doi.org/10.1007/s12665-016-5444-5.
27. Kirgiz, M. S. Smart *Nanoconcretes and Cement-based Materials: Nano Size Particle Packing for Nanoconcretes and Cement-based Materials: Mathematical Models, Theory, and Technology* (Mohd Shahir Liev, Phuong Nguyen-Tri, Tuan Anh Nguyen, Saeid Kakooei, Eds.), Elsevier, ISBN: 978-0-12-817854-6. https://doi.org/10.1016/B978-0-12-817854-6.00003-9.
28. Papayianni, I., Pachta, V., and Stefanidou, M. (2013). Analysis of ancient mortars and design of compatible repair mortars: The case study of Odeion of the archaeological site of Dion. *Construction and Building Materials*, **40**, 84–92.
29. Mohammed, A., Sanjayan, J. G., Duan, W. H., and Nazari, A. (2015). Incorporating graphene oxide in cement composites: A study of transport properties. *Construction and Building Materials*, **84**, 341–347.

Chapter 5

Transforming Waste Oil Sand Ash into Construction Material Using Net-Zero Energy

M. Hesham El Naggar,[a] Mehmet Serkan Kirgiz,[b] André Gustavo de Sousa Galdino,[c] Roberto A. Gonzalez-Lezcano,[d] Raul D. S. G. Campilho,[e] and Konstantinos G. Kolovos[f]

[a]*Department of Civil and Environmental Engineering, Western University, London, Ontario N6A 5B9, Canada*
[b]*Academia of ACMCEN, Beaverton, OR 97075, USA*
[c]*Federal Institute of Education, Sciences and Technology of Espírito Santo, Av. Vitória, 1729, Jucutuquara, Vitória, ES, 29040-780, Brazil*
[d]*Architecture and Design Department, Escuela Politécnica Superior, Universidad CEU San Pablo, CEU Universities, 28040 Madrid, Spain*
[e]*INEGI – Pólo FEUP, Rua Dr. Roberto Frias, 400, 4200-465 Porto, Portugal*
[f]*Department of Military Sciences, Division of Physical Sciences and Applications, Hellenic Army Academy, Vari, Greece*
nakres42@yahoo.com

The aim of the chapter is to measure the potential of transforming waste oil sand ash into construction materials using net-zero energy.

Handbook of Innovative Adhesive Technology: Biodegradable and Cost-Effective Binder for Built Environment and Engineering Sciences
Edited by Mehmet Serkan Kirgiz

ISBN 978-981-5129-51-9 (Hardcover), 978-1-003-59892-3 (eBook)
www.jennystanford.com

For this purpose, waste oil sand ash was collected from a manufacturer in Canada. In order to test the effectiveness of the proposed method, waste oil sand ash was collected from a manufacturer in Canada. In order to demonstrate the efficacy of the tested methods, particle size distribution, scanning electron microscopy (SEM) observation, energy dispersive spectroscopy (EDS), and X-ray diffraction (XRD) were employed. Furthermore, comprehensive information about the production technique of oil sand waste (OSW) and its economic value was included in the study. Additionally, to extract bitumen from oil sand, a new method containing a distillation process was suggested. It is possible to conclude that OSW can be grinded until it is finer than 3 μm and also that OSW contains about 85% silicon dioxide (SiO_2) and displays various geometric as well as amorphous shapes. Moreover, it is possible to utilize waste oil sand ash in making construction materials using net-zero energy in terms of its fineness, content of SiO_2, and various geometric and amorphous shapes.

5.1 Introduction

Oil plays a dominant role in meeting the world's energy needs and will continue to do so for decades to come. Even with the investments that Canada and other countries have made in renewable energy, energy efficiency, and other measures to support a low-carbon economy, according to the International Energy Agency's *World Energy Outlook 2012*, the world's oil demand is expected to grow by more than 0.5% per year until 2035 and the global economy will continue to rely more on oil than on any other fuel. Canada's oil sands, a bitumen-soaked sand deposit, encompass more than 149,000 km^2 of Alberta in three regions: Athabasca, Cold Lake, and Peace River. The "in place" volume of the oil sand resource is 1.84 trillion barrels, significantly more than the oil that has been produced in the world to date. The cumulative production since the beginning of the oil sand industry in Canada until 2016 is less than 7% of the current established reserves [1]. Figure 5.1 illustrates the number of barrels of fuel oil obtained from oil sands by various countries.

Here are some significant statistics—initial volume in place: 1.84 trillion barrels; initial established reserves: 176.8 billion barrels; remaining established reserves (2016): 165.4 billion barrels; and cumulative production (2016): 11.4 billion barrels. Figure 5.2 shows the situation of fuel oil production from oil sands in Canada.

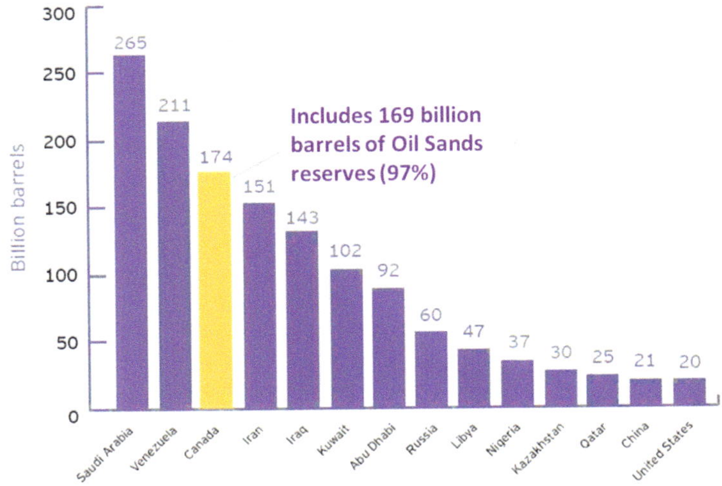

Figure 5.1 Barrels of fuel oil obtained from oil sands by various countries.

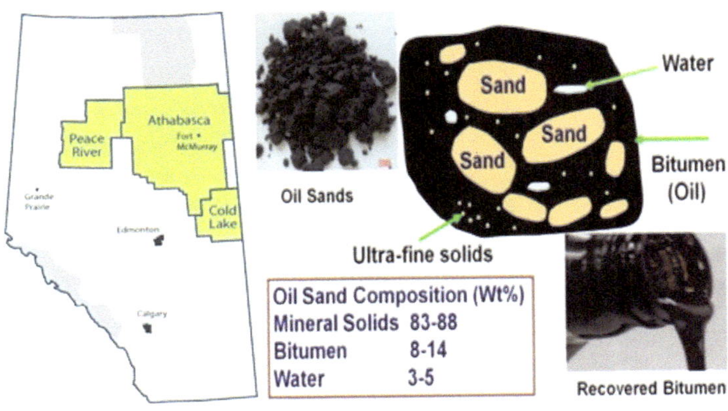

Figure 5.2 Situation of fuel oil production from oil sands in Canada.

Oil sand production methods include drilling (in situ steam-assisted gravity drainage [SAGD]) and surface mining. Manufacturers use SADG when the deposit of oil sand is deeper than 70 m below the ground surface. On the contrary, the surface mining method is used when the deposit of oil sand is less than 70 m below the ground surface. Oil sands are recovered using mining and drilling (in situ) depending on how deep the reserves are deposited. About 80% of oil sand reserves are too deep to be mined and are thus recovered in place, or in situ, by drilling wells. Drilling (in situ) methods create minimal land disturbance and do not require tailings ponds. Oil sands that lie more than 70 m (200 feet) below the ground are recovered using drilling methods. Advanced technology is used to inject steam, combustion, or other sources of heat into the reservoir to warm the bitumen so that it can be pumped to the surface through recovery wells. Around 20% of the oil sand reserves are close enough to the surface to be mined using large shovels and trucks [1]. Figure 5.3 illustrates how fuel oil is obtained from oil sand through drilling, mining, and tailing ponds.

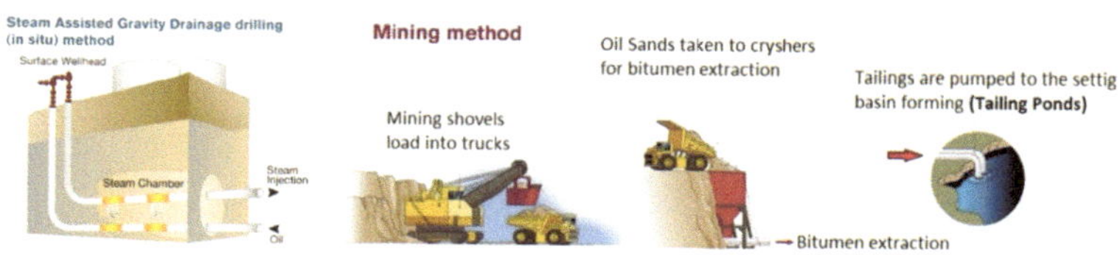

Figure 5.3 Fuel oil made from oil sands through drilling, mining, and tailing ponds.

Figure 5.4 shows recyclable waste in fuel oil manufacturing from oil sands.

Figure 5.4 Recyclable waste in fuel oil manufacturing from oil sands.

Mining operations produce about 52% of OSW. The drilling cutting waste is one of the most difficult challenges for the oil sand industry in terms of waste management. Usually, tailings are disposed in toxic tailing ponds. The migration of these pollutants into the groundwater system can adversely affect our environment.

5.2 Application of Treated OSW in Controlled Low Strength Materials

Figure 5.5 shows various applications of recyclable waste from oil sand ash in the construction industry. The current chapter aims at recycling and reusing treated OSW (TOSW) and reducing the amount of solid waste disposal and contributing to the effort of reducing the footprint of the oil sand industry on the environment as well as reducing the use of expensive fuel ash in controlled low strength material (CLSM) mixture.

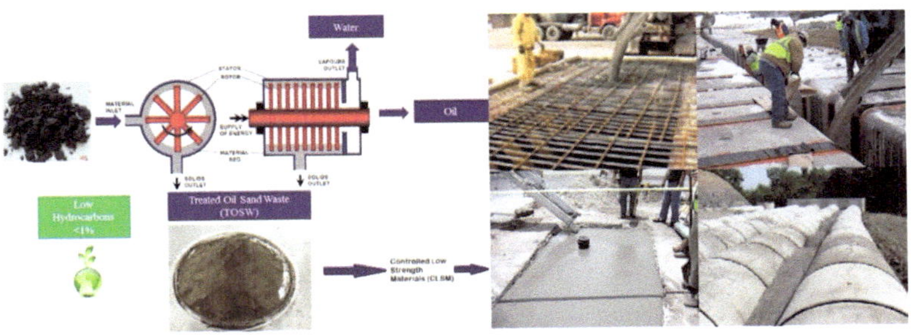

Figure 5.5 Various applications of recyclable wastes from oil sand ash in the construction industry.

Flowable self-leveling cementitious materials used as an alternative to compacted cementitious materials may be utilized in geotechnical applications. CLSM is usually a mixture of cement, sand, water, and fly ash (Fig. 5.5).

5.3 Materials and Methods

This chapter presents a working plan in two phases. The first phase is the evaluation of the effect of TOSW on the fresh and hardened properties of CLSMs. The second phase is the evaluation of the dynamic properties of CLSMs incorporating TOSW. To measure the influence of TOSW on the properties of materials, the chapter determines its descriptive properties using XRD, SEM, and EDS and evaluates its particle size distribution.

5.4 Results and Discussion

5.4.1 XRD, SEM, and EDS of TOSW

The most significant physical property of TOSW is the presence of a very fine quartz powder. Figure 5.6 shows the real image of TOSW, SEM micrographs, and EDS analysis of TOSW [1].

Figure 5.6 Real image of TOSW, XRD, SEM micrograph, and EDS analysis of TOSW.

The microstructure of TOSW is presented in Fig. 5.6 by various geometric and amorphous shapes, as well as SiO_2, which can give stability to the paste structure of construction materials. There is a visible densification around TOSW grains due to the high content of SiO_2, which can convert CaOH to a fibrous calcium silicate hydrate (CSH) gel in cement paste. It can be understood from the SEM imaging in Fig. 5.6 that a high degree of hydration can occur in cement paste with the supplement of TOSW. Several hydration compounds spotted during the EDS analysis also supported that they were silicate, since they exhibited an Si/Ca ratio of about 0.95 [1]. In comparison to other supplements with TOSW, the greatest compressive strength can be obtained by materials containing TOSW (Fig. 5.6).

5.4.2 Particle Size Distribution of TOSW

Figure 5.7 shows the particle size distribution of both TOSW and conventional Portland cement. The oil sand waste (OSW) is a significant source of oil production, with the majority of this oil destined for the Canadian market. Yet, there is no value-added economic profit in its reuse, recycle, and upcycling. Since there are heat and pressure in the occurrence of the oil sand in the ground layers and it has mineral solids between 83% and 88%, its waste has the strong possibility of showing latent hydraulic activity, similar to marble powder, on materials such as cement, concrete, mortar, lime, and stucco. Thus, it has much value than we think. Regarding the production technique of oil in Canada, which is the third largest oil producer and has the third largest oil deposit, including 169 billion barrels of oil sand reserves (97% of the total reserves), in the world, there are two important methods. One of them is the drilling method, either in situ method or SAGD.

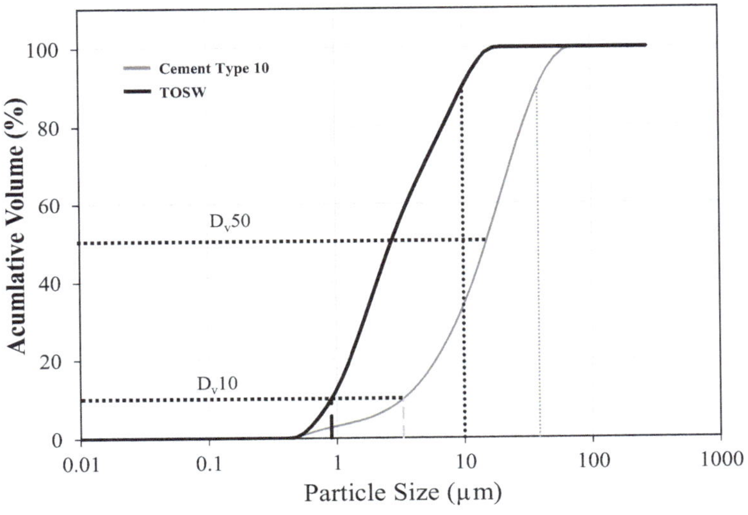

Figure 5.7 Particle size distribution of TOSW and conventional Portland cement.

Additionally, SAGD is used when the oil is deeper than 70 m under the ground surface. The second method, known as surface mining, is useful for drawing out oil when the deposit is not deeper than 70 m under the ground surface. After oil production, there are ponds that are toxic and contain the residue of oil sand. The material can be reused and recycled into the economy, and it can be upcycled through new waste disposal methods. For instance, its fine powder can be replaced with cement in the manufacturing of cement-based materials, and its coarse and fine particles can be replaced with aggregate in the manufacturing of cement-based materials. Additionally, its fine powder can be blended in the mixing of fresh cement-based materials as the latent hydraulic additive material, and it can be used as a raw material in the binder manufacturing directly. The simplest method is selling OSW as raw materials to cement manufacturers after extracting its bitumen. Moreover, the particle size of TOSW is finer than that of type 10 cement. The fineness of type 10 cement is 15.09 μm, while that of TOSW is 2.71 μm (Fig. 5.7) [1].

5.5 Conclusion

Treated OSW can be grinded until it is finer than 3 μm. TOSW also contains about 85% SiO_2 and appears in various geometric as well as amorphous shapes. Moreover, waste oil sand ash can be utilized in the making of construction materials using net-zero energy in terms of its fineness, SiO_2 content, and various geometric and amorphous shapes.

Reference

1. El Naggar, M. H. 2019. *Cementitious Materials for Geotechnical Applications Incorporating Oil Sand Waste.* Proceedings book of the Third Annual International Conference on Eco-Sustainable Construction Materials. ISBN 978-605-031-179-2.

Chapter 6

Ecological, Sustainable, Green Brick Materials: Unfired Manufacturing System, Some Features, and Usage for Greener Built Environment and Construction Technology

John Kinuthia,[a] Mehmet Serkan Kirgiz,[b] André Gustavo de Sousa Galdino,[c] Raul D. S. G. Campilho,[d] Konstantinos G. Kolovos[e]
[a]*Advanced Materials Testing Centre (AMTeC), School of Engineering, University of South Wales, United Kingdom*
[b]*Academia of ACMCEN, Beaverton, OR 97075, USA*
[c]*Federal Institute of Education, Sciences and Technology of Espírito Santo, Av. Vitória, 1729, Jucutuquara, Vitória, ES, 29040-780, Brazil*
[d]*INEGI – Pólo FEUP, Rua Dr. Roberto Frias, 400, 4200-465 Porto, Portugal*
[e]*Department of Military Sciences, Division of Physical Sciences and Applications, Hellenic Army Academy, Vari, Greece*
nakres42@yahoo.com

The manufacture of steel and Portland cement consumes an enormous quantity of raw materials. The manufacturing processes

Handbook of Innovative Adhesive Technology: Biodegradable and Cost-Effective Binder for Built Environment and Engineering Sciences
Edited by Mehmet Serkan Kirgiz

ISBN 978-981-5129-51-9 (Hardcover), 978-1-003-59892-3 (eBook)
www.jennystanford.com

involved are associated with high-energy consumption and unsightly excavations, especially when left without attention to landscaping, and exacerbate atmospheric pollution. Portland cement forms the main binder material in concrete and represents the costliest constituent in concrete and masonry. Therefore, the environmental and economic impacts of Portland cement are huge. It is responsible for about 10% of the global human-derived carbon dioxide emissions. For these reasons, the possibility of replacing, or minimizing the usage of, traditional building and construction materials with low-cost greener materials from nature, industry, or agriculture will yield significant benefits, and numerous researches have been undertaken on different waste materials. Research, so far, is incomplete in this direction, and more coverage is required. Pending issues relate to durability and versatility as a building material to resolve issues related to domestic fixtures (shelves, etc.), jointing, demolition, and/or reuse among others. Most of the major waste and by-product streams have been explored, leaving only marginal and waste materials. This chapter explores these issues with the aim of enhancing the uptake of earth-based raw materials from nature, industry, or agricultural sources.

6.1 Introduction

The use of clay-based materials has a long and prehistoric background. For example, in 1997, Ei-Naggara *et al.* demonstrated the advantages of soil reinforcement. The study evaluated the mechanical properties of soil. However, it also highlighted the potential of clay-based materials for fundamental construction practice. This is due to the increased cost of construction, reduced sustainability, and reduced reliability of indoor quality [1]. Many researchers followed the study of Ei-Naggara *et al.* Furthermore, today constructors and end users can find several advanced studies on using nano clay in various construction composite materials [2]. Panda *et al.*, in 2019, carried out a study on using nano-attapulgite clay (NC) in fly ash mixture to improve the composite material's printability in 3D printing. They revealed that NC increased the static yield stress of composite materials without significantly affecting the apparent viscosity because the particles of NC undergo reflocculation and enhance thixotropy [3]. The NC increases viscosity recovery and develops the structural build-up at various resting periods and shear ratios. The authors posit that these developments have been beneficial for 3D composite printing in the construction industry. In this context, the composite can be worked upon layer by layer without any damage [3].

Raj *et al.*, in 2018, conducted comprehensive research on silane composite functionalized with nano clay and polypropylene fibers. The authors found that the addition of nano clay allows proper development of the properties of silane, which can provide better interfacial adhesion for both nano clay and polypropylene, and their proper dispersion. The authors proposed that the resulting silane composite, which contains nano clay and polypropylene, could be utilized by various industries and in a wide range of applications, including the automotive and construction industries [4]. He *et al.*, in 2019, worked on a method of preparing rubber composite–compounded nano clay gel and revealed that although a high-performance rubber-clay nanocomposite could be achieved, the preparation method of the nano rubber composite was still limited. The most significant result of their study was that the tensile strength of rubber-clay nanocomposites was more than 250% higher than that of pure rubber composites [5]. Adamu *et al.*, in 2019, reported a study on the effects of montmorillonite-based nano clay and polyvinyl alcohol (PVA) treatments for the physical, mechanical, and thermal properties of bamboo-reinforced nanocomposites. This study revealed that compared to standard bamboo, the modulus of elasticity (MOE) increased from 7.82 GPa to 17.32 GPa, and the modulus of rupture (MOR) increased from 68.67 MPa to 118.74 MPa in bamboo-reinforced nanocomposites. The results of the differential scanning calorimeter (DSC) and thermogravimetric analysis of bamboo-reinforced nanocomposites and standard bamboo indicated that the former exhibited superior thermal properties compared to the latter [6].

Mohan and Kanny, in 2019, published a study on the strength features of unmodified composites and nano clay–added banana fibers reinforced with epoxy binder composites. They compared the compressive, tensile, interfacial, and energy absorption features with those of unmodified composites and nano clay–added banana fibers reinforced with epoxy binder composites. The authors observed that the incorporation of nano clay into banana fiber composites resulted in notable improvements in compressive yield stress, modulus, and strength, with values that were 17%, nearly two times, and 43% higher, respectively, when compared with the banana fiber-reinforced epoxy composite that did not contain any nano clay. They suggested that the nano-modification technique was so beneficial that it could be applied to various natural fibers to realize the potential application of various composites [7]. Xia *et al.*, in 2019, studied the effect of nanoparticle shape on the mechanical properties of polymer-matrix composites containing two types of polyethylene glycol (PEG)-inserted silica nanoparticles—nanoplatelets and nanospheres—with similar surface features prepared through biomimetic synthesis. Their comparative analysis demonstrated that the two nanoparticles exhibited different behaviors during the shear yielding of composite, leading to different ways of energy dissipation and toughening. In addition, the creep behaviors and dynamic mechanical properties were closely related to the shape of nanoparticles as well [8]. Ayatollahi *et al.*, in 2019, worked on the fracture analysis of rubber nano clay composites. They developed a hyperelastic model to predict the fracture feature of rubber nano clay composites [9]. Moustafa *et al.*, in 2019, wrote a review on the future vision and challenges of eco-friendly polymer composites for green packing. Their study provided insights for all those who would work on eco-friendly composite materials. The study's results pulled scientific attention on renewable and biodegradable bio-composite materials in a broad field of practices, such as smart packing, biomedicine and drug delivery, biomembranes, and automobiles as well as industrial composting. The corresponding author obtained various biomass from renewable resources, for instance coffee ground, nanocellulose, and date stones to develop intelligent reinforcing agents in biodegradable biopolymers for improving their overall properties [10].

Although there is plenty of information in the previous paragraph, the association of clay building materials with mud huts, as depicted in Fig. 6.1, is correct but not sufficient. There has been serious rethinking and modernizing, and clay-based materials can now produce good quality final products. Figure 6.2 shows researched clay products by the authors, suggesting good workmanship. Full characterization also suggests good performance [11–13]. It is also possible to achieve aesthetically pleasant construction, either using individual products such as those shown in Fig. 6.2, or using other techniques appropriate for earth construction. Good workmanship can be achieved for both individual and large-scale housing projects using unfired clay materials, in both developed and developing countries (Figs. 6.3 and 6.4).

Figure 6.1 Example association of clay building materials with mud huts.

Figure 6.2 Researched clay product by the authors, suggesting good workmanship.

There is a need to explore and highlight sustainability in researched soil-based engineering materials. There are diverse sustainability issues related to all soil-based materials in the construction industry, which need the formulation of standards and codes of practice [14]. Most importantly, this significant material resource should capitalize and highlight unfired soil systems for enhanced eco-friendly replacement of traditional binder materials such as cement and steel-based materials, whose manufacture, as already narrated, is energy intensive, consumes large volumes of precious natural raw materials, and contributes to significant atmospheric emissions.

Figure 6.3 Example of individual aesthetically pleasing construction using rammed earth technique. (Courtesy: Centre for Alternative Technology (CAT) at Machynlleth, Wales, UK)

Figure 6.4 Stabilized earth block buildings in India. (Courtesy: B. V. Venkatarama Reddy, Centre for Sustainable Technologies, Indian Institute of Science (IISc) Bangalore, India)

Fired soil-based materials are also attractive options of construction materials, especially compared with concrete and steel, but their use should also be accompanied by a consideration of the firing mechanisms and likely costs and future impact. In general, all building and construction materials should endeavor to address the key aspects of manufacturing, design, utilization, maintenance, and/or whole life cycle analysis among other issues. It is strongly recommended that considerations include details of material sources, potential future supplies, appreciable raw material and final product characterization, and fully detailed methodologies. In the assessment of performance, it is imperative that there is provision for the degree of impact such materials are likely to have on the community, government, and industry, and of the foreseeable futuristic developments in their utilization. Going forward into the future, it is important that a positive reflection of earth-based materials is created, with a balanced view of the negative aspects. In the past, the negative aspects of low strength, volume instability, and low esteem in their use have tended to predominate. However, recently, the trend is changing when considerations for climate change have become global focal issues. This gradual development has somewhat turned the tide in favor of earth-based construction, minimizing negative and inconsequential outcomes associated with their use.

Over the last few decades, clay-based materials have started to gain recognition and the respect they once had in the past. This is credited to clay-based material enthusiasts who have brought to the fore the well-recognized and indisputable benefits of these materials. These benefits include low impact on the environment, particularly if the materials are not fired, and their ability to passively control interior humidity. The materials need not be associated with mud huts only, and aesthetically pleasant buildings can be made as already demonstrated [15, 16].

The downside of earth-based materials includes the bad reputation these materials have, especially when compared with the apparent advantages of energy-intensive materials such as Portland cement and, to some extent, lime. Another notable drawback is the often slow speed of construction and the sensitivity clay materials sometimes have when one has to balance cost and immediate- and long-term performance. While some approaches may require little or no prior experience, others such as the modern and beautiful appearances shown in Figs. 6.3 and 6.4 require extra care during construction in order to maintain material mixing regimes and the resultant uniformity in density, strength, and color [17–19]. When the material is used in individual compacted units/bricks, a detailed characterization of the product is necessary, and thus rigorous laboratory testing regimes are essential [20, 21]. Some products can be innovative, such as the (tongue and grove) interlocking earth block shown in Fig. 6.5.

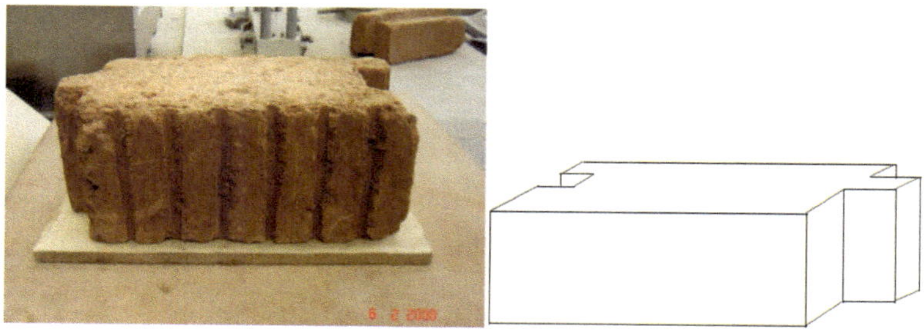

Figure 6.5 An innovative tongue and grove interlocking earth block.

Research work by the authors, in association with a community project in Africa, fully characterized the earth block shown in Fig. 6.5. The block weighed 7.55 kg and could support a load of 217.3 kN over an area of 364 cm^2. This suggested a carrying capacity of 6.0 N/mm^2. This is significant strength for a stabilized soil block. The edges were not very good, and considering that these would be rectified in service by the use of an appropriate mortar, the actual load-bearing capacity is, therefore, likely to be much higher.

The typical strength values for good, stabilized soil materials compacted in the laboratory fall in the range of 2–4 N/mm^2, after being cured for at least 28 days. The higher strength for the block in Fig. 6.5 was attributed to:

- Good compaction during production using a fabricated steel frame, rather than a laboratory kit.
- Good bearing capacity of the target material—laterite (or murram as is commonly known).

A complete analysis of the block material using thermogravimetric and derivative thermogravimetric (TG/DTG) analyses (Fig. 6.6) suggested that the lateritic block material consisted of high clay content. This is shown by the significant peak within the temperature peak between 500 and 600°C. This is the region where clay mineral generally loses its combined hydroxyl (OH–) component.

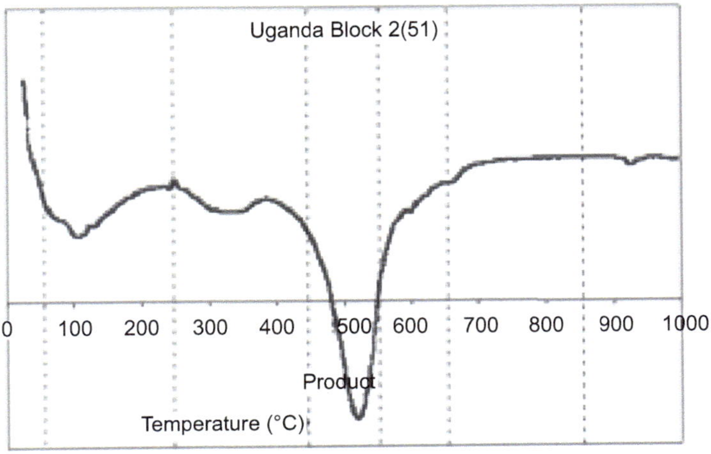

Figure 6.6 Thermogravimetric analysis suggesting high clay content in the lateritic clay soil used in the soil blocks.

Typical concrete building blocks usually target about 7 MPa as characteristic (28-day) strength. Thus, this block is reasonably strong and suggests the presence of a significant amount of binder or binding properties. This, in turn, suggests that:

- Possible savings could be made by reducing the cement content used.
- There was a need to review its use to match the strength with requirements.
- It was possible to produce two or more products of different strength values and hence different applications.

The local community in Africa was very innovative and incorporated further innovation to make curvy blocks (Fig. 6.7). The curved nature enabled versatile applications such as in curved walls for curvy buildings, for making water tanks, and possibly other applications. This curved block weighed 6.72 kg and could only support a load of 38.6 kN over an area of 138 cm^2. This suggested a carrying capacity of 2.8 MPa. The block was difficult to load, as there was no proper loading surface. The curved nature of the block exacerbated loading procedures. This shape of the block contributed to the low load-bearing capacity value observed, when individual units were tested. It is feasible that in normal service, when the interlocking units are used in a wall or panel, the load-bearing capacity is bound to be much higher than observed in the laboratory.

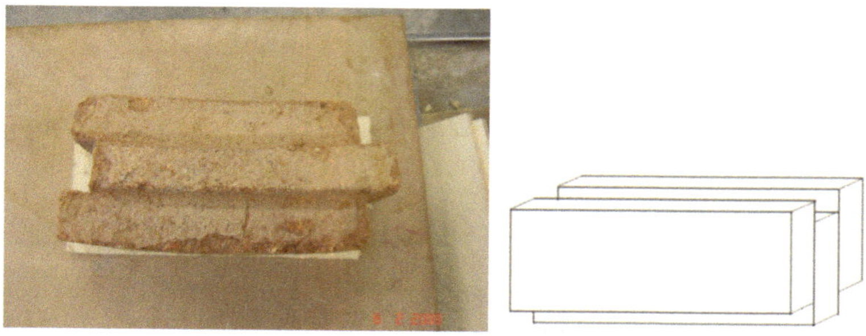

Figure 6.7 Examples of innovative curvy and interlocking earth block for curvilinear construction.

The blocks failed in shear (Fig. 6.8), suggesting that further strengthening against this mode of failure need to be addressed. The strengthening measures will depend on the use of the blocks, for example in water tanks or houses.

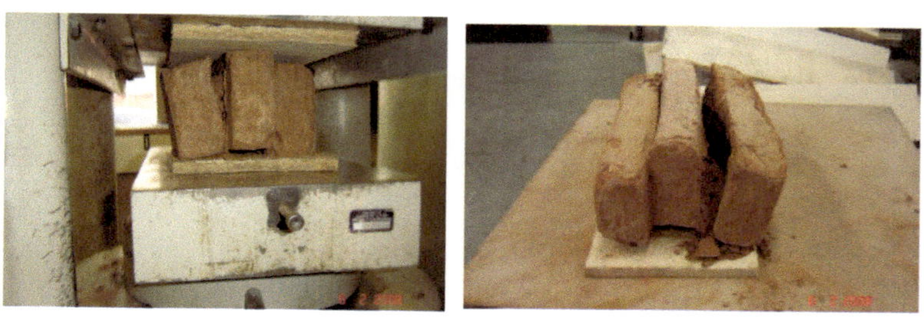

Figure 6.8 Rigorous testing for full characterization of innovative earth blocks.

Practical experience suggests that strength tests should involve some element of filling using an appropriate mortar to prepare for loading. This would better simulate the final resting place of the blocks in service, with better resultant working strength values. For both jointed and interlocking building units, it is far more practical to test complete building elements made from individual block units, such as walls or panels, as shown in Fig. 6.9.

Figure 6.9 Rigorous testing using complete elements such as walls or panels for the full characterization of innovative earth blocks.

6.2 Durability

It is appropriate now to shift emphasis toward the durability of compressed earth materials. In the fired clay category, the material is mixed with a significant amount of water and either extruded or molded in a timber, plastic, or steel mold into a regular shape. There are reports of sand molds also being used [16]. Before firing, the cast or extruded bricks or blocks are allowed to condition or dry slightly to reduce the water content, so as to reduce the shrinkage on firing.

Firing of clay drives out not only the free water used in the mixing process but also any water that is chemically combined or adsorbed in the interlayers of the clay microstructure [22]. The firing also goes further to dehydroxylate (removal of the –OH group) and/or decarboxylate (removal of CO_2) any carbonates present and sinters the materials (if temperature is high enough, typically >1000°C) to a very robust fired product. The loss of free and chemically bound water reduces the swelling potential associated with clay soils. Considering that some soils can exhibit excessive swelling potential as high as 2000%, the firing process is very effective in imparting volume stability to a target clay soil material [22]. The strong bonding reduces the weakening effects of porosity and imparts chemical resistance. Firing is, however, expensive due to its high energy intensity and represents about 85% of the energy involved in the manufacture of fired clay materials [23]. This is a major problem in both developed and developing countries alike.

The unfired systems, compared to their fired counterparts, have a relatively lower energy consumption due to the absence of the firing process. It is, however, best to start by addressing the target raw material itself—clay soil. The most basic historic clay-based materials capitalize on inherent clay properties, in particular, its cohesive nature. Simple techniques have merely involved the selection of clay deposits. It is possible to find significant deposits of clay without having to remove any coarse materials. However, clay is a versatile material, and this makes it an attractive construction material. It is able to include a significant amount of coarse materials while remaining plastic. Traditional houses have been built with clayey silt, clayey sand, and boulder clay.

In its wet plastic form, clay can be molded into a regular shape, usually but not limited to a cuboid form. The regular shape is especially necessary if the wet material is to be sun dried for later use. Alternatively, the wet material can be placed either in a panel formwork, as in rammed earth, or in a cage-like open formwork, as in traditional huts. The wet cohesive materials stick together and form a solid irregular mass. The structure can be used with or without further finishing, such as plastering. If well plastered, it is hard to believe that a structure is made of irregular mud bulbs.

6.3 Industrial Additives

Soils can be strengthened by other unfired processes, such as soil-stabilization techniques. When soils are stabilized with lime and/or Portland cement, a colloidal product predominantly composed of a calcium silicate hydrate (CSH) gel is formed, although aluminum phases and traces of iron may also be present. The complex gel gradually changes with time by partially crystallizing, resulting in strength gain in a mechanism very similar to that in Portland cement hydration. The composition of this colloidal product is dependent on material ingredients used: the compounds of calcium, silicon, aluminum, and traces of iron coming from the lime or Portland cement used as a stabilizer; aluminum and silicon from the soil; and finally the water added for the stabilization process. This colloidal $CaO–Al_2O_3–SiO_2–H_2O$ system is beneficial for strength development, although it is prone to ingress of water and other elements. The quantity and long-term development of this colloidal product influence the total porosity and affect strength and volume stability. The minimization of pores is obviously beneficial to strength in construction materials in general [24, 25]. When well protected, all goes well, and the materials develop significant strength to enable applications in roads and foundations

6.3.1 Ground Granulated Blast Furnace Slag

This is an industrial by-product material that results from the manufacture of steel from iron ore in a blast furnace (Fig. 6.10). The material has successfully been applied in the concrete industry, where it results in a reduced use of Portland cement, an energy-intensive material with a significant negative environmental impact.

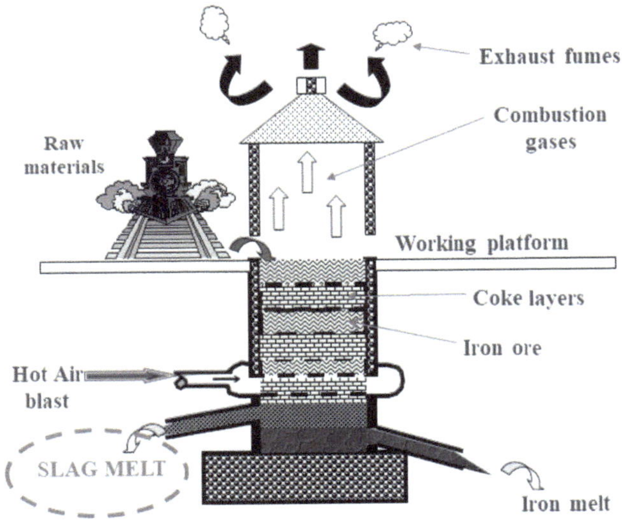

Figure 6.10 Schematic drawing showing the manufacture of GGBS.

The use of ground granulated blast furnace slag (GGBS) also results in improved durability in concrete. The material has had very little impact in masonry until in recent times but is now becoming common for applications in soil stabilization [21]. Figure 6.11 shows that higher values of compressive strength at 7 days of curing are being obtained using formulations containing increasing amounts of GGBS. Please note that GGBS on its own resulted in no strength development. The test samples [26] also contained increasing amounts of sulfur (introduced during the addition of gypsum) to simulate a sulfate-bearing soil and were moist cured at 30°C and 100% relative humidity.

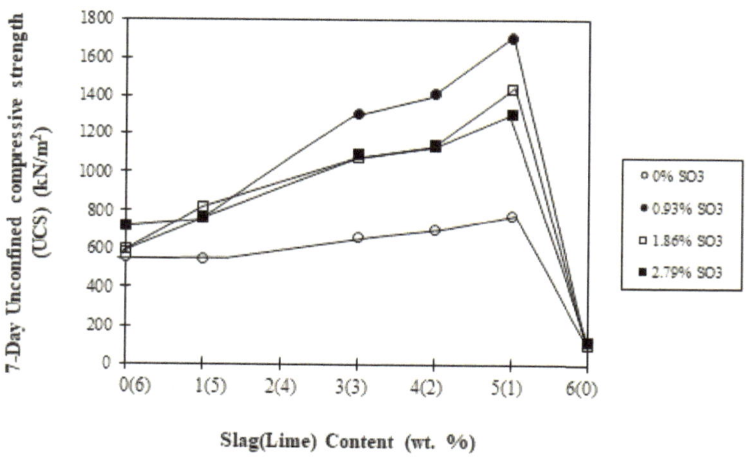

Figure 6.11 Plots of 7-day-cured cylinders of kaolinite clay, stabilized using a reduced amount of lime (and increasing GGBS).

6.3.2 Pulverized Fuel Ash

For many years, coal has been a dominant source of energy worldwide. The waste from this industry ranges from unusable mining debris, collectively referred to in various terms such as coal waste, colliery spoil, colliery waste, coal mining waste, coal mine tailings, and possibly other terms, to the usable pulverized fuel ash (PFA), which results from the burning of coal as a fuel. The burned waste is predominantly the fine particulate material collected from the flue gasses, mainly by electrostatic precipitation. It is commonly referred to as fly ash (FA) in America and other places or as pulverized fuel ash (PFA) in the United Kingdom and some parts of Europe and beyond. There is also the relatively coarser waste referred to as bottom ash (BA). All these ash variables can be used together with clay soil to make sustainable earth-based materials [27].

6.3.3 Waste-Paper Sludge Ash

Most paper is coated with clay and limestone to create smooth surfaces to write on. When the paper is then recycled, the clay and the limestone remain in the sludge from which the recyclable cellulose fibrous material is removed during recycling. When this sludge is combusted to reduce the volume of waste going to a landfill, the clay and limestone are heated in the process, undergoing more or less a similar heating process as Portland cement, albeit at lower temperatures. This material can be used successfully as a soil stabilizer [28].

6.3.4 Agricultural Additives

For most countries, many major industrial activities are directly or indirectly related to agriculture. For these countries, any major breakthrough in the development of sustainable infrastructure cannot afford to ignore waste from the agricultural sector. The author has been brainstorming on this agenda, in liaison with like-minded co-workers for some time. The initial focus has been on activities that produce large agricultural waste streams, including the growing of palm oil and sugarcane [20], rice [28], and fibrous materials for fiber-based soil–cement blocks—bagasse, straw, kenaf, bamboo, jute, coir, and durian [29]. In Cameroon, inspired researchers at the *Mission de Promotion des Materiaux Local* (MIPROMALO)—a local materials promotion authority based in Yaoundé—have provided valuable collaborative support on rice research [28].

6.3.5 Construction Using Clay-Based Systems

The increased use of clay-based materials in building and construction has challenging practical implications, relative to cement-based materials, at all stages of their use. Undertaking research in clay-based materials is also challenging, due to low strength and variability in other performance characteristics. There are also equal if not more difficult challenges to execute research findings into practice, at various stages of infrastructure development—materials manufacture.

6.3.5.1 Storage and construction

The storage of unfired clay products is also critical, as heavy losses can be incurred if the material is left unprotected during storage. For this reason, unfired clay materials, without exception, are routinely protected in some manner during storage, no matter how ineffective the protection might be. Experience with large-scale factory-type production of unfired clay-based materials is rare and recent. During automated production, the control of compaction moisture content, the handle ability of the freshly compacted product, the care needed during early transport to storage/curing yard, and the subsequent and continued care during loading on trucks all pose problems. The significant product losses during transport and construction have all discouraged large-scale application of earth-based materials.

6.4 Concluding Remarks and Future Trends

There is an observable increase in the researched use of compressed earth masonry units. These units are increasingly being seen as industrialized materials and no longer considered as only appropriate or applicable to traditional approaches for self-construction [30]. Many examples exist of the increasing use of industrial and agricultural waste as well as by-product materials in compressed earth masonry, and it is not possible to cover or quote them all here. A brief survey of the materials, techniques, and applications of today's masonry suggests a thriving ongoing research and innovation-led development of compressed earth masonry technology. There is also a gradual shift from the traditional fired clay masonry categories to unfired categories. While the fired categories adopted a blanket approach of firing clay irrespective of clay composition, there is a far wider variety of material selection and technology adopted for the unfired masonry category. This is because the durability imparted by the firing process is harder to mimic or achieve in unfired systems, resulting in the pursuance of alternative approaches to materials usage to include industrial and agricultural waste and by-product materials. There is also increasing use of synthetic (industrial) and natural fibres for reinforcement, chemical cementation, attention to particle matrix configurations, and variable and careful selection of applicability scenarios.

The pursuance of the rapid setting and rapid strength development by way of geopolymer routes to cementation is a new development that is likely to gain interest among researchers in this subject. Geopolymer soil systems will obviate the use of calcium-based cementation, which has been the tradition and has been associated with not only atmospheric pollution but also slow pozzolanic cementation. There are teething issues with geopolymer cementation, some systems needing elevation of temperature to achieve meaningful results. Geopolymers in general also require expensive chemicals to achieve alkaline activation. Notwithstanding these drawbacks, there are drivers that may capture researchers' interest, such as readily available sources of precursor high-silica and/or high-alumina-content industrial by-products such as fly ash, slag, and silica fume among others, as well as agricultural wastes such as rice husk ash [28]. Fortunately, as mentioned earlier, there is also a gradual shift in emphasis from a blanket stipulation of standard properties and material expectations, relying rather on recommendations by the manufacturer for the appropriate use and for the possible consequences of lack of adherence, by the end user, to these recommendations. The future will, therefore, most likely witness enhanced end-user sensitivity to global changes in materials, technology and material failure mechanisms, and better or closer liaison with material manufacturers, coupled with an awareness of relevant government policies, guidance, and legislation. This development or trend is likely to enhance reliance on compressed earth masonry, and it is hoped that mortgage lenders and insurance industries will be quick to respond to the fast-changing landscape, resulting in an ever-increasing adaptation of compressed earth masonry in future social and commercial housing. This scenario suggests an urgent need for a near-universal harmonization, consensus building and characterization, and/or classification of compressed earth–based building and construction materials, considering the wide range of materials and variable manufacturing methodologies likely to be encountered. In terms of further research work needed, tests on durability need to be well delineated, with those for fired systems adopting considerations that suit fired systems separate from those best suited for unfired systems.

Unfired systems will, in the future, need to be tested and categorized in specialized ways, with special key differences highlighted between stabilized and non-stabilized compressed earth systems. Both test categories should embrace and embed the indicative tests of compressive strength and water absorption. The more severe cyclic freeze–thaw and wet–dry tests should be addressed in reference to stabilized systems, while the water spray and drip test methodologies addressed in reference to unstabilized systems. Also as future development, there is a need to capitalize on and strengthen the assessment of the environmental benefits of clay-based materials, as well as their positive whole life cycle considerations [14, 27]. This is not an easy task, especially for the life cycle analysis. The input parameters for a robust whole life cycle assessment are not easy to identify in their entirety. It is also not easy to assess them equitably for different manufacturing locations and situations. In this regard, pioneering work has elucidated the challenges faced by researchers in their attempts to assess the whole life cycle of clay-based individual elements [14]. The work suggests that there is a need to go beyond the current emphasis on energy consumption and atmospheric carbon dioxide emission. Diversity in the whole life cycle assessment is needed to cover a wider range of parameters.

References

1. Ei-Naggara, M. E., Kennedy, J. B., and Ibrahima, E. M. Mechanical properties of reinforced soil. *Composites Part B: Engineering*, **28**(3), 275–286, 1997.
2. Rajaei, M., Kim, N. K., Bickerton, S., and Bhattacharyya, D. A comparative study on effects of natural and synthesised nano-clays on the fire and mechanical properties of epoxy composites. *Composites Part B: Engineering*, **165**, 65–74, 2019.
3. Panda, B., Ruan, S., Cise, U., and Tan, M. J. Improving the 3D printability of high volume fly ash mixtures via the use of nano attapulgite clay. *Composites Part B: Engineering*, **165**, 75–83, 2019.

4. Raji, M., El Mehdi Mekhzoum, M., Rodrigue, D., Qaiss, A. E. K., and Bouhfid, R. Effect of silane functionalization on properties of polypropylene/clay nanocomposites. *Composites Part B: Engineering*, **146**, 106–115, 2018.

5. He, S., He, T., Wang, J., Wu, X., Xue, Y., Zhang, L., and Lin, J. A novel method to prepare acrylonitrile-butadiene rubber/clay nanocomposites by compounding with clay gel. *Composites Part B: Engineering*, **167**, 356–361, 2019.

6. Adamu, M., Rahman, M. R., and Hamdan, S. Formulation optimization and characterization of bamboo/polyvinyl alcohol/clay nanocomposite by response surface methodology. *Composites Part B: Engineering*, **176**, https://doi.org/10.1016/j.compositesb.2019.107297, 2019.

7. Mohan, T. P., and Kanny, K. Compressive characteristics of unmodified and nanoclay treated banana fiber reinforced epoxy composite cylinders. *Composites Part B: Engineering*, **169**, 118–125, 2019.

8. Xia, S., Liu, X., Wang, J., Kan, Z., Chen, H., Fu, W., and Li, Z. Role of poly(ethylene glycol) grafted silica nanoparticle shape in toughened PLA-matrix nanocomposites. *Composites Part B: Engineering*, **168**, 398–405, 2019.

9. Ayatollahi, M. R., Heydari-Meybodi, M., Berto, F., and Yahya, M. Y. Mixed-mode fracture in EPDM/SBR/nanoclay rubber composites: An experimental and theoretical investigation. *Composites Part B: Engineering*, **176**, 2019, https://doi.org/10.1016/j.compositesb.2019.107312.

10. Moustafa, H., Youssef, A. M., Darwish, N. A., and Abou-Kandil, A. I. Eco-friendly polymer composites for green packaging: Future vision and challenges. *Composites Part B: Engineering*, **172**, 16–25, 2019.

11. Oti, J. E., and Kinuthia, J. M. Stabilised unfired clay bricks for environmental and sustainable use. *Applied Clay Science*, **58**, 52–59, 2012.

12. Oti, J. E., Kinuthia, J. M., and Bai, J. Developing unfired stabilized building materials in the UK. *Engineering Sustainability*, **161**, 211–218, 2008.

13. Oti, J. E., Kinuthia, J. M., and Bai, J. Using slag for unfired-clay masonry-bricks. *Construction Materials*, **161**, 147–155, 2008.

14. Marcelino-Sadaba, S., Kinuthia, J., Oti, J., and Meneses, A. S. Challenges in life cycle assessment (LCA) of stabilised clay-based construction materials. *Applied Clay Science*, **144**, 121–130, 2017. https://doi.org/10.1016/j.clay.2017.05.012.

15. Pacheco-Torgal, F., Lourenco, P. B., Labrincha, J. A., Kumar, S., and Chidaprasirt, P. The durability of compressed earth-based masonry bricks, in *Eco-Efficient Masonry Bricks and Blocks: Design, Properties and Durability*, 2015.

16. Sutton, A., Walker, P., and Black, D. *Unfired Clay Masonry: An Introduction to Low-Impact Building Materials*. Information Paper IP 16/11 BRE-University of Bath, 2011.

17. Kinuthia, J. M., and Oti, J. E. 2012. Designed non-fired clay mixes for sustainable and low carbon use. *Applied Clay Science*, **59/60**, 131–139, 2012.

18. McGregor, F., Heath, A., Shea, A., and Lawrence, M. The moisture buffering capacity of unfired clay masonry. *Building and Environment*, **82**, 599–607, 2014.

19. Miqueleiz, L., Ramirez, F., Seco, A., Nidzam, R. M., Kinuthia, J. M., Abu Tair, A., and Garcia, A. The use of stabilised Spanish clay soil for sustainable construction material. *Engineering Geology*, **133–134**, 9–15, 2012.

20. Mofor, L., Kinuthia, J. M., Melo, U. C., and Djialli, D. Energy in Cameroon: The potential for transporting waste to riches. *Material Solution*, **1**, 18–21, 2009.

21. Nidzam, R. M., and Kinuthia, J. M. Sustainable soil stabilisation with blast furnace slag (GGBS): A review. *Construction Materials*, **163**, 157–165, 2010.

22. Grim, R. E. *Clay Mineralogy*, 2nd edn, McGraw Hill Book Company, 1968.

23. Heath, A., Walker, P., Fourie, C., and Lawrence, M. Compressive strength of extruded unfired clay masonry units. *Ice Proceedings, Construction Materials*, **162**, 105–112, (CM3), 2009.

24. Benavente, D., Garcıa, M. A., Fort, R., and Ordoñez, S. Durability estimation of porous building stones from pore structure and strength. *Engineering Geology*, **74**, 113–127, 2004.

25. Molina, E., Cultrone, G., Sebastian, E., Alonso, F. J., Carrizo, L., Gisbert, J., and Buj, O. The pore system of sedimentary rocks as a key factor in the durability of building materials. *Engineering Geology*, **118**, 110–121, 2011.

26. Wild, S., Kinuthia, J. M., Jones, G. I., and Higgins, D. D. Suppression of swelling associated with ettringite formation in lime-stabilised sulphate-bearing clay soils by partial substitution of lime with ground granulated blast furnace slag (GGBS). *Engineering Geology*, **51**, 257–277, 1999.

27. Nidzam, M., and Kinuthia, J. Strength and environmental evaluation of stabilised Clay-PFA eco-friendly bricks. *Construction Building Materials*, **125**, 964–973, 2016.
28. Nidzam, R. M., and Kinuthia, J. M. Effects of mellowing sulfate-bearing clay soil stabilized with wastepaper sludge ash for road construction. *Engineering Geology*, **117**, 170–179, 2011.
29. Khedari, J., Watsanasathaporn, P., and Hirunlabh, J. Development of fibre-based soil-cement block with low thermal conductivity. *Cement Concrete Composites*, **27**, 111–116, 2005.
30. Cid-Falceto, J. C., Mazarron, F. R., and Cañas, I. Assessment of compressed earth blocks made in Spain: International durability tests. *Construction Building Materials*, **37**, 738–745, 2012.

Chapter 7

Adhesively Bonded *T*-Joint Optimization by Cohesive Models: Effect of Adhered Material, Adhesive Type, Geometry, and Bonding Technique

J. P. M. Lopes,[a] Raul D. S. G. Campilho,[a,b] and Mehmet Serkan Kirgiz[c]
[a]*ISEP – School of Engineering, Polytechnic of Porto, Rua Dr. António Bernardino de Almeida, 431, 4200-072 Porto, Portugal*
[b]*INEGI – Pólo FEUP, Rua Dr. Roberto Frias, s/n, 4200-465 Porto, Portugal*
[c]*Academia of ACMCEN, Beaverton, OR 97075, USA*
nakres42@yahoo.com

Different bonded joint configurations are available to designers. While single-lap joints (SLJs) are the most common in application and research, double-lap joints, scarf joints, and *T*-joints find specific applications. *T*-joints are seldom studied in the literature, but these are used, for instance, in aircraft to bond the stiffener beams to the skin, or in cars between the *B*-pillar and the rocker. Due to the high

Handbook of Innovative Adhesive Technology: Biodegradable and Cost-Effective Binder for Built Environment and Engineering Sciences
Edited by Mehmet Serkan Kirgiz

ISBN 978-981-5129-51-9 (Hardcover), 978-1-003-59892-3 (eBook)
www.jennystanford.com

stress concentrations, *T*-joints often fail under average stresses much lower than the adhesive strengths, giving rise to the necessity for proper design and strength improvement methodologies. This chapter initially aims to validate the cohesive zone modeling (CZM) technique with experiments and then use it to numerically evaluate and optimize the performance of *T*-joints subjected to peel loads. CZM is nowadays regarded as the most powerful strength prediction tool for adhesive joints and can be a valuable tool to improve *T*-joints. Different features are addressed for a complete analysis: adhesive type, geometrical parameters, dual-adhesive technique for strength improvement, and composite joints. The evaluated geometrical parameters include the base adherend thickness (*a*), *T*-part thickness (*t*), overlap or bonding length (*l*), and curvature radius (*r*). As a result of this work, the model was successfully validated, and clear design guidelines were provided to define the ideal geometric and material (adhesive) conditions for best performance.

7.1 Introduction

Currently, polymers are commonly used as adhesive materials in different fields of knowledge, and adhesive joint design is a hot topic in the academia. In comparison with traditional joining methods such as bolting or welding, adhesive bonding technology has several advantages, namely improved stress fields, capacity to join different material types without compromising their integrity (e.g., absence of welding and hole drilling in composites), vibration reduction, and sealing [1]. Limitations of adhesive bonding include the necessity for surface treatment before the application of the adhesive, typical disassembly difficulties, limited temperature and humidity, and appearance of undesired peel (σ_y) stresses [2]. Among varying joint designs that a designer can choose from, the most simple and common for application and research purposes is the SLJ [3]. Other joining strategies consist of double-lap, scarf, stepped-lap, *T*, and tubular joints. The *T*-joint configuration finds application in different scenarios, including aviation, to bond stiffeners to the aircraft skin, and automotive industry, for the pillar-to-rocker connection. A common *T*-joint design joins a base adherend to the *T*-part along a certain bonding length. The *T*-joint's main purpose is the transfer of peel loads between the base panel and the web [4]. Irrespective of the bonded joint configuration, techniques are available in the literature for strength improvement, leading to improved load transfer and efficiency, by reducing stress concentrations and increasing the average stress at failure, including finding the optimal *l*, *a*, and the adhesive thickness (t_a) [5]; applying adhesive fillets [6] and chamfers [7]; and using the dual-adhesive technique [8]. The bi- or dual-adhesive technique is an invention of Raphael [9], and it consists of combining a low-stiffness adhesive at the highly stressed overlap edges, with a high-stiffness adhesive at the inner overlap, typically under low stresses, aiming to make stress distributions more uniform and thus increase the joint strength.

To enable the design and application of adhesive bonding technology in different applications, it is necessary to accurately predict the adhesive joints' strength, which highly facilitates tailoring a specific solution to a given application without extensive experimental testing. Strength prediction of adhesive joints began with analytical models, for example the work of Hart-Smith [10], which were proposed to evaluate the joint performance. However, these techniques failed to work for novel ductile and tough adhesives. Finite element method (FEM) analyses, on the other hand, find accuracy for complex geometrical and material problems. With this approach, it becomes possible to account for virtually all possible effects and constraints necessary for the accurate assessment of the joints' behavior. Continuum mechanics is the simplest approach—by considering the maximum stresses, strains, or strain energy given by the models (or analytical approach), and comparing with the mechanical properties of adhesive to predict failure [11]. Fracture mechanics assumes that structures and adhesive layers can have discontinuities and cracks, and that these can grow if specific conditions are met. CZM is a powerful approach for describing damage growth along the cohesive process zone leading to the total detachment of the constituent parts of a structure. By CZM, damage can propagate along predefined paths established by the modeler. However, CZM relies on accurately estimated cohesive strengths in tension and shear (t_n^0 and t_s^0, respectively), and fracture toughnesses in modes I and II (G_{IC} and G_{IIC}, respectively). The extended finite element method (XFEM) is a more recent addition to the pool of available methods [12], additionally enabling the propagation of cracks along arbitrary paths, unlike CZM. By XFEM, the mesh does not need to match the discontinuities, nor it is necessary to re-mesh the model in the crack vicinity.

Adhesively bonded *T*-joints are a less common topic of research in the literature, compared to simpler geometries, but a few authors have addressed this geometry. The work of Zhan *et al.* [13] used damage mechanics to evaluate different *T*-joint configurations under tensile loads. The joint materials consisted of 2060-T8 aluminum alloy and the two-part epoxy adhesive EA9394. The experimental tests carried out positively validated the proposed numerical approach. The maximum load (P_m) increased for higher bonding areas, and the horizontal bondline was more effective in the contribution to P_m than the vertical bondline. In fact, by considering a vertical bonding in a *T*-joint, the increase in P_m was negligible, although the bondline increased by 123%. In a different work [14], CZM was used to address the adhesive type and geometry effects on the bending characteristics of *T*-joints. The joints were made of carbon-epoxy (IM7/8552) and glass-epoxy (E-glass/Epon 828) adherends. The joint strength depended on the applied adhesive and gave the following results (from best to worst): Araldite® AV138, Araldite® 2015, SikaForce® 7752, Cytec FM73, and Cytec FM300. *T*-joints with sharper corners led to the highest stresses at the respective vicinity and smaller stresses at the overlap ends, while rounded corners promoted the opposite behavior. Higher stringer lengths diminished von Mises stresses at the bond ends, although the corner stresses increased. Increasing the web and stringer thickness increases von Mises stresses in the adhesive. Ramezani *et al.* [15] experimentally analyzed the dual-adhesive technique in SLJ by employing digital image correlation (DIC). The analysis comprised a flexible and a stiff adhesive, and different geometries, by varying l, a, and t_a. DIC clarified the influence of these conditions on strain distributions, showing that by increasing the applied load, the maximum compressive strains tend to approach the adhesives' interface. Compared to the single-adhesive joint, the dual-adhesive revealed higher strength.

This chapter initially aims to validate the CZM technique with experiments and then use it to numerically evaluate and optimize the performance of *T*-joints subjected to peel loads. CZM is nowadays regarded as the most powerful strength prediction tool for adhesive joints and can be a valuable tool to improve *T*-joints. Different features are addressed for a complete analysis: adhesive type, geometrical parameters, dual-adhesive technique for strength improvement, and composite joints.

7.2 Materials and Methods

7.2.1 Description of *T*-Joint Models

The *T*-joint geometry is shown in Fig. 7.1, and it consists of a double-*L* arrangement forming a *T*, using two bent parts bonded in their length. The *T*-joint includes a material-free portion at the radius region, between the double-*L* parts. A peel displacement was applied at the *T*-part edge, while the base (flat) adherend was clamped at both edges. The base dimensions are the following (all dimensions are given in mm): total length l_t = 200; *T*-part height l_a = 40; l = 30; a = 3; t = 1, 2, 3, and 4; r = 6; t_A = 0.2; and width b = 25. On the other hand, the effect of geometric parameters is assessed through the comparative evaluation to the base geometry. The following configurations are tested (dimensions in mm): a (1, 2, 3, and 4), t (0.5, 1, 1.5, 2, and 2.5), l (10, 20, 30, and 40), and r (3, 6, 9, and 12).

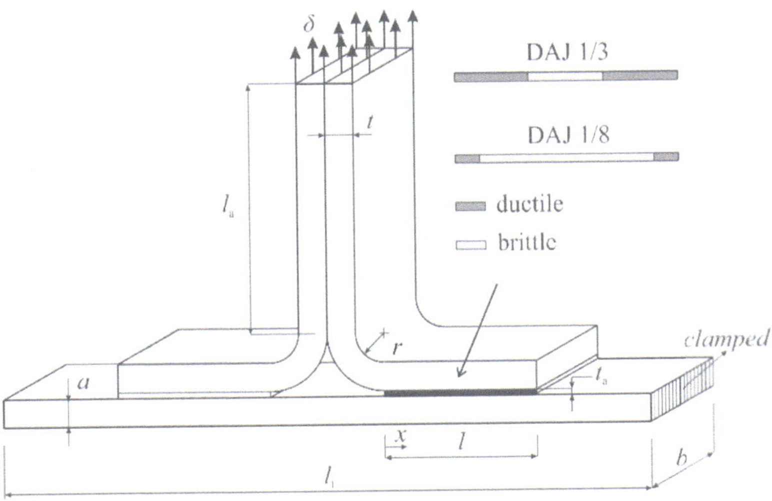

Figure 7.1 *T*-joint geometry.

7.2.2 Adherends and Adhesives

The *T*-joint analysis was mainly conducted with AW 6082-T651 aluminum alloy adherends (base adherend and *T*-part), with carbon–epoxy adherends being analyzed in the second stage. This aluminum alloy was previously tested [16], leading to the mechanical properties: Young's modulus (E) of 70.07 ± 0.83 GPa, tensile yield stress (σ_y) of 261.67 ± 7.65 MPa, tensile strength (σ_f) of 324 ± 0.16 MPa, and tensile failure strain (ε_f) of 21.70 ± 4.24%. On the other hand, the carbon–epoxy adherends are made of unidirectional prepreg plates with carbon fiber reinforcement and epoxy matrix (SEAL® Texipreg HS 160 RM; Legnano, Italy). The ply unit thickness of the prepreg is 0.125 mm, and manual fabrication consisted of hand lay-up and

Table 7.1 Properties of the adhesives Araldite® AV138 [18], Araldite® 2015 [19], and SikaForce® 7752 [20]

Property	AV138	2015	7752
Young's modulus, E [GPa]	4.89 ± 0.81	1.85 ± 0.21	0.49 ± 0.09
Poisson's ratio, ν	0.35[a]	0.33[a]	0.30[a]
Tensile yield stress, σ_y [MPa]	36.49 ± 2.47	12.63 ± 0.61	3.24 ± 0.48
Tensile failure strength, σ_f [MPa]	39.45 ± 3.18	21.63 ± 1.61	11.48 ± 0.25
Tensile failure strain, ε_f [%]	1.21 ± 0.10	4.77 ± 0.15	19.18 ± 1.40
Shear modulus, G [GPa]	1.56 ± 0.01	0.56 ± 0.21	0.19 ± 0.01
Shear yield stress, τ_y [MPa]	25.1 ± 0.33	14.6 ± 1.3	5.16 ± 1.14
Shear failure strength, τ_f [MPa]	30.2 ± 0.40	17.9 ± 1.8	10.17 ± 0.64
Shear failure strain, γ_f [%]	7.8 ± 0.7	43.9 ± 3.4	54.82 ± 6.38
Toughness in tension, G_{IC} [N/mm]	0.20[b]	0.43 ± 0.02	2.36 ± 0.17
Toughness in shear, G_{IIC} [N/mm]	0.38[b]	4.70 ± 0.34	5.41 ± 0.47

[a] Manufacturer's data
[b] Estimated by an inverse method

curing in a press for 1 h at 130°C and 2 bar pressure. The elastic orthotropic data for a single unidirectional lamina, required for the numerical models, are described in a previous work [17]. The work is mainly conducted with the ductile epoxy adhesive Araldite® 2015, but a total of three adhesives were used, namely, for the adhesive type and dual-adhesive analyses. Araldite® AV138 is a brittle epoxy adhesive, and SikaForce® 7752 is a ductile polyurethane adhesive. The collected properties are described in Table 7.1. All properties were defined in previous research works, providing the necessary information for CZM modeling. The tensile bulk data (E, σ_y, σ_f, and ε_f) resulted from bulk tests on dog bone specimens. The shear properties were acquired using thick-adherend shear tests (TASTs) on bonded steel specimens. The tensile and shear fracture data were found by performing double-cantilever beam and end-notched flexure tests, respectively.

7.2.3 Model Pre-processing

CZM was applied to infer the *T*-joint peel performance, after validation with experimental results, taking advantage of the Abaqus® software and the built-in CZM module. Only half model was considered, by applying symmetry conditions, aiming to reduce the computational load associated to the static analyses carried out for the different joint configurations. The models were built as two dimensional (2D) due to the constant shape along the width, and the geometry partitioning technique facilitates the mesh construction by reducing element distortions. The aluminum and carbon–epoxy adherend partitions were populated by solid elements with four nodes and plane–strain formulation (CPE4 elements) [21]. The thin adhesive layer was modeled using a single row of CZM elements (COH2D4 elements with four nodes). Figure 7.2 shows a mesh example.

A cohesive-type section was applied to the adhesive layer, with geometry-defined thickness. All aluminum alloy and carbon–epoxy sections were considered solid and homogeneous. Application of the boundary conditions involved clamping the base adherend edge and vertically pulling the *T*-part with a prescribed displacement at its top edge. Vertical symmetry conditions were applied in all models. To simulate decohesion, a triangular CZM was applied in the adhesive layer, which included a mixed-mode formulation involving a quadratic stress criterion to assess damage initiation and a linear power-law energetic criterion to deal with crack propagation. The detailed description of the CZM model is given in Ref. [22]. The data on adhesives for the simulations are provided in Table 7.1.

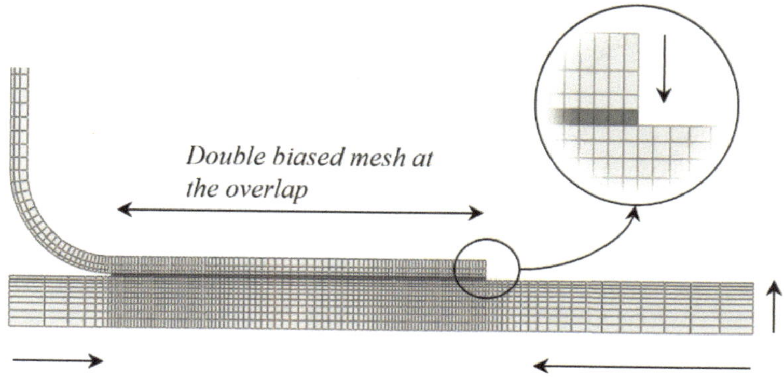

Figure 7.2 Mesh refinement for the *T*-joint with $t = 1$ mm.

7.3 Results

The results section is divided into CZM validation, to enable the numerical application of the procedure to different analyses, followed by a P_m analysis on different *T*-joint features: adhesive type (AV138, 2015, and 7752), geometrical parameters (a, t, l, and r), dual-adhesive technique, and composite adherends.

7.3.1 CZM Validation

Initially, CZM validation is carried out with the *T*-joint geometry depicted in Fig. 7.1, aluminum alloy adherends, and the 2015. For fabrication, the adherends were initially cut and bent into the *L*-shape. Surface preparation for bonding involved grit blasting and cleaning with acetone. The adhesive was then poured on the surfaces to join, and the assembly was cured in a jig, which assured the alignment of joint. The desired t_a was assured with calibrated steel spacers. Pressure was applied individually to each specimen with grips [23].

The joints were cured over a 1-week period and, finally, the excess adhesive resulting from the fabrication process was removed by trimming. Testing was accomplished at least 1 week after assembly and executed at room temperature and a velocity of 1 mm/min to emulate static test conditions, in a Shimadzu AG-X 100 machine. Five repetitions were considered for each joint configuration, defined by t (Fig. 7.1). Visual observation of the failed specimens showed cohesive failures of the adhesive for all t values and repetitions. Figure 7.3 summarizes the collected experimental P_m data and the respective numerical predictions.

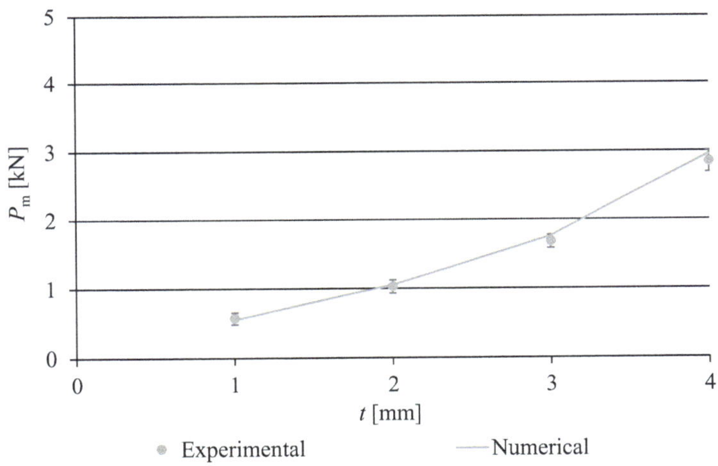

Figure 7.3 Experimental and numerical P_m versus t data.

An analysis of the experimental P_m reveals a clear t effect with the benefit of higher t, especially between 3 mm and 4 mm. Considering the validation adhesive (2015) and compared to the base geometry with t = 1 mm, the percentile P_m improvements for consecutive increases in t values up to 4 mm are 81.2%, 197.8%, and 403.7%, thus showing a significant effect of the T-part stiffness on the joint performance. Previous analyses showed that this difference is due to the reduction in σ_y and shear (τ_{xy}) peak stresses [24]. Additionally, this adhesive can plasticize before failure, which enables a larger region of stress transfer by increasing t. The CZM P_m predictions showed a good match to the experiment data, although P_m was always predicted by excess of the experimental average values, except for 1 mm (4.3% difference by default). For t = 2, 3, and 4 mm, the experimental P_m was below the numerical predictions by 2.2%, 4.5%, and 3.5%, in this order. As a result of the validation process, the CZM technique can be considered reliable for the following numerical parametric analysis.

7.3.2 Adhesive Type

The initial analyzed effect on the *T*-joints' P_m behavior is the adhesive type, considering the three adhesives described in Section 7.2.2. For this and the subsequent analyses, the P_m data are collected from the load–displacement (P–δ) curves resulting from the numerical simulations, which is also used to analyze stress distributions and failure modes. Figure 7.4 shows that a higher l improves P_m in *T*-joints

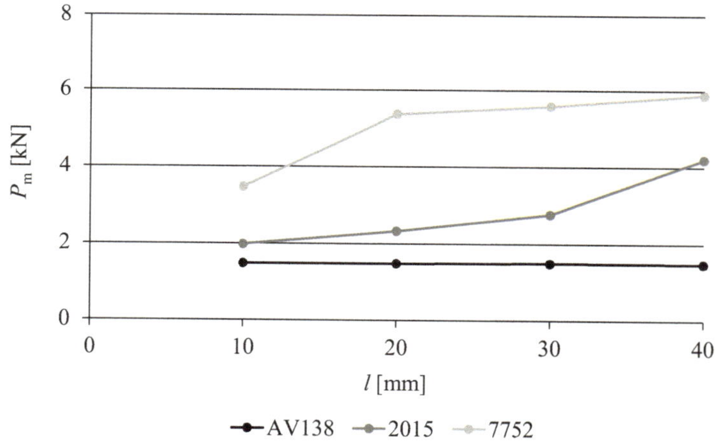

Figure 7.4 Adhesive type effect on P_m.

bonded with ductile adhesives, i.e., 2015 and 7752. On the other hand, this improvement was not possible by using AV138, due to its inherent brittleness, and no sensible variations in P_m are detected in the plot. For this adhesive, P_m diminishes from 1473 N to 1471 N between l = 10 mm and 40 mm. As a totally opposite behavior, the *T*-joints bonded with the 7752, which is the most ductile of the three adhesives, and suffered a P_m improvement from 3452 N to 5878 N between l = 10 mm and 40 mm (70.3% difference). It is possible to conclude from this analysis that ductile adhesives accomplish significant P_m improvements, since these adhesives can absorb and plasticize at the loci of peak stresses before reaching failure, thus providing a major advantage over stronger but brittle adhesives, which in this geometry do not work well due to the predominant peel load.

7.3.3 Geometrical Parameters

Different geometrical parameters are also addressed for the base adhesive (2015), supported by the presented P_m data in Fig. 7.5. Each parameter is addressed individually:

- The parameter a highly affects P_m, with a benefit of higher a (Fig. 7.5a). Between a = 1 mm and 4 mm, the increase in P_m was significant (by 112.8%), and it is justified by the reduction in the deflection of the base, which makes stress distributions more uniform. In the opposite direction, higher a values reduce the allowable displacement of the joints before failure. On the other hand, an increasing a decreases the failure displacement.
- A higher t has an identical effect to a, i.e., P_m increases with t (Fig. 7.5b). A 30.8% P_m improvement was found by increasing t from 0.5 to 2.5 mm. Nonetheless, for t = 1.5 mm, a short disruption to this tendency was found. A numerical analysis showed that σ_y and shear (τ_{xy}) stresses increased at the bond edges and that the failure mode changed by comparing t = 0.5 and t = 1 mm with t = 1.5, 2, and 2.5 mm, leading to inconsistent P_m evolution. In the first scenario, failure took place at the curvature radius region, while in the second scenario, it occurred at the bond edges. The failure displacement slightly increased for a higher t.
- P_m steadily increases with l, with preponderance for higher l (Fig. 7.5c). The relative improvement between limit l was 112.2%. A detailed analysis showed that, for this adhesive, a higher l enables plasticization at the stress concentration regions, leading to P_m improvements due to higher average stresses and higher bonded areas. The failure displacement increased with l as well.

- The r effect is depicted in Fig. 7.5(d), showing a linear increase in P_m by reducing r (difference of 17.2% between the limit r values). It was found that this variation was caused by alterations in σ_y and τ_{xy} stresses at the radius vicinity. The displacement at joint failure also decreased for a higher r.

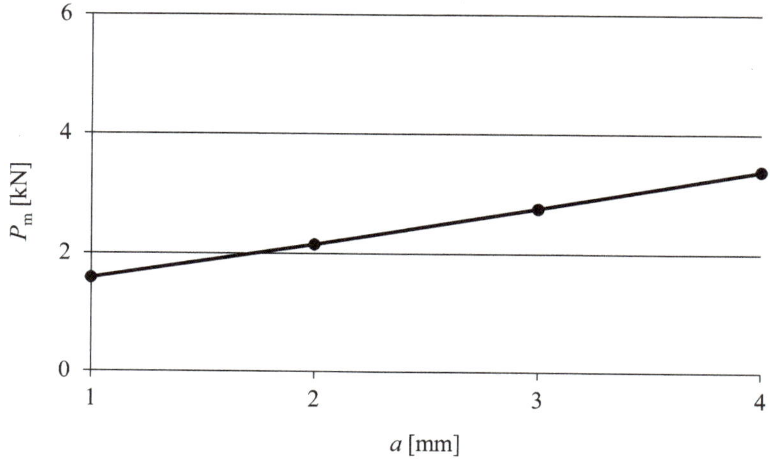

(a)

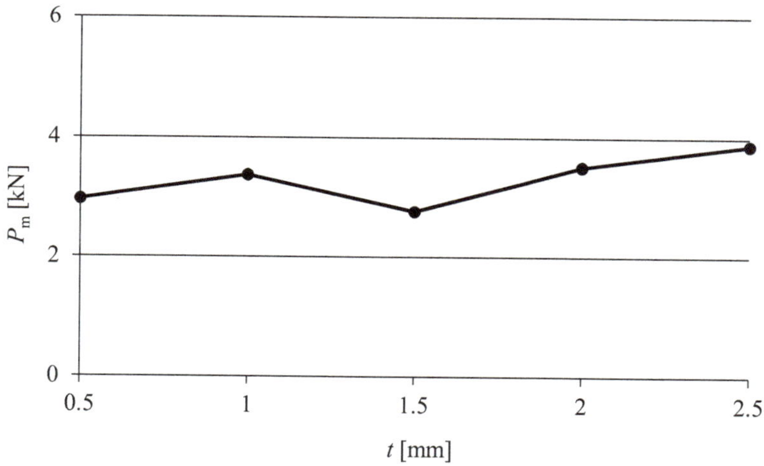

(b)

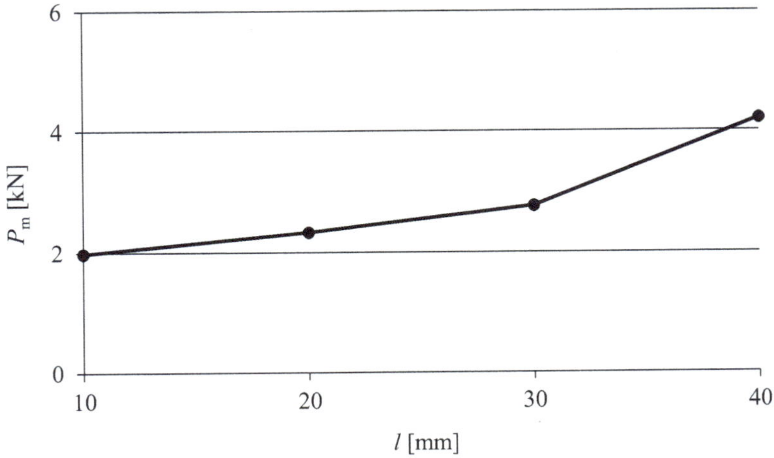

(c)

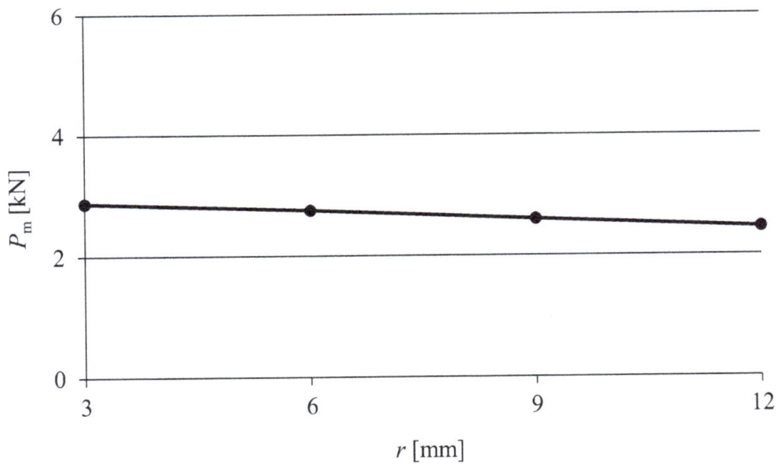

(d)

Figure 7.5 Geometrical parameter effect on P_m: influence of a (a), t (b), l (c), and r (d).

7.3.4 Dual Adhesives

Figure 7.6 presents the P_m-t data considering the 2015, 7752, and 7752–2025–7752 (dual-adhesive) joints. The dual-adhesive joint can perform better than a single-adhesive joint in the T-joint configuration, but the joint geometry should be carefully chosen. Of all tested joints, those bonded with the 2015 provide the most modest P_m results. Thus, the associated P_m data points are the benchmark of the complete data pool. Between the two adhesives considered for this analysis, the 2015 is stronger than the 7752 (Table 7.1). However, the 7752 ductility largely exceeds that of the 2015. Since *T*-joints have a predominant peel load, and significant peak stresses are present in the adhesive layer, the 7752 is able to sustain higher loads before failure, thus leading to improved P_m. Table 7.2 quantifies the differences between all joints and evaluates ΔP_m (variation in P_m to the single-adhesive *T*-joint bonded with the 2015). The *T*-joint bonded with the 7752 manages to provide ΔP_m up to 38%, for $t = 1$ mm, which corroborates the former discussion. The advantage of the dual adhesive joint technique varies according to *t*, and also to the adhesives' proportion. Considering the dual-adhesive joint 1/8, real advantage over the 7752 is restricted to $t = 3$ mm, in which case ΔP_m reaches 20%. On the other hand, the dual-adhesive joint 1/3 clearly performs best between all tested conditions (single- or dual-adhesive), even though, for $t = 1$ mm, ΔP_m is identical to the single-adhesive joint with the 7752.

Between all *t*, the highest ΔP_m corresponded to $t = 1$ mm and reached 38%. ΔP_m values were less significant for the remaining *t*, but the variations in the single-adhesive joints bonded with the 2015 and 7752 cannot be disregarded as well. Comparison of the best solution between all *t* shows that this variable is relevant for joint design, and the best P_m result corresponds to $t = 3$ mm.

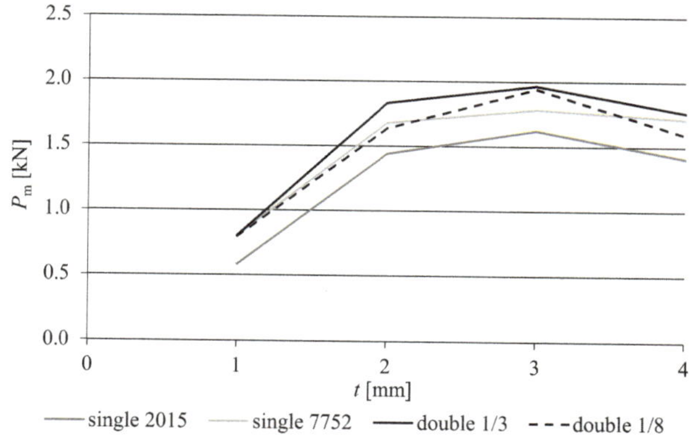

Figure 7.6 Dual-adhesive effect on P_m.

Table 7.2 Collected P_m and ΔP_m data for the dual-adhesive joint analysis

t (mm) Adhesive	1	2	3	4
2015	582.65 N	1437.4 N	1621.86 N	1407.29 N
7752	803.91 N (38%)	1678.74 N (17%)	1782.81 N (10%)	1712.64 N (22%)
Dual-adhesive joint 1/8	796.18 N (37%)	1636.25 N (14%)	1943.46 N (20%)	1592.09 N (13%)
Dual-adhesive joint 1/3	803.94 N (38%)	1832.58 N (27%)	1968.68 N (21%)	1764.87 N (25%)

7.3.5 Composite Adherends

The *T*-joint geometry depicted in Fig. 7.1 is addressed as a function of *l*. Although different failure paths were considered in the numerical models (including composite failures), typically fracture was cohesive in the adhesive. The numerical P_m-l data are provided in Fig. 7.7. It is found that P_m increases with *l* from 10 to 20 mm, although a further increase in *l* leads to a strength depreciation. The relative variations in P_m, compared against the base geometry with l = 10 mm, were calculated as +30.5%, +9.7%, and −20.1% for increasing *l* up to 40 mm. A detailed analysis of the software model showed that σ_y and τ_{xy} peak stresses at the bond edges tend to increase with *l*, due to the increased stiffness of the bonded set, which may cancel the theoretical advantage in having a bigger area of adhesive by increasing *l*. In the current case, it was found that the P_m disruption has its origin in a modification of the failure mode. Up to l = 20 mm failures were cohesive in the adhesive layer, while above l = 20 mm, the radius region failed prematurely to cohesive failure of the adhesive, obstructing higher P_m and leading to the reported behavior. Moreover, and differently to what is known to occur for SLJ, for instance, this failure mode alteration effectively reduced P_m over a geometry with less bonded area.

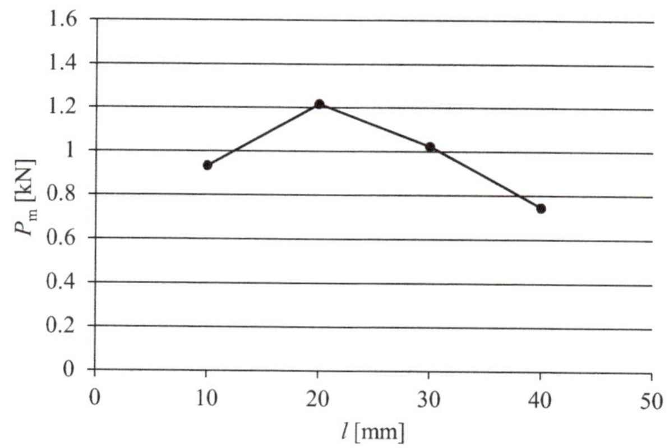

Figure 7.7 Material effect on P_m.

7.4 Conclusion

The present work was accomplished to optimize aluminum alloy *T*-joints under peel loads, using CZM, after validation with experiments. The analysis comprised different adhesives, geometric variables (*a*, *t*, *l*, and *r*), dual-adhesive technique, and composite adherends. The numerical analysis was carried out in Abaqus®. Validation was initially positively done, which enabled the subsequent numerical approach. The following conclusions are divided into these four approaches:

- **Adhesive type:** The 7752 provided the highest P_m disregarding *l*, showing a major improvement up to l = 20 mm. This behavior is justified by the plasticization allowance at the adhesive edges, even though the adhesive is the least strong between the three adhesives. Applying AV138 showed no advantage to higher *l* because of the loading nature. The 2015 has intermediate properties and showed a smaller P_m evolution than the 7752.
- **Geometrical parameters:**
    - A higher *a* linearly increased P_m on account of resulting in a stiffer base and less deflection. This parameter was found to be highly relevant for the joint behavior.
    - An increasing *t* promotes a P_m improvement in the considered range, apart from the *T*-joint with t = 1.5 mm due to the modification of the failure mode.

  - An increasing l provides a higher bonded area, which increases P_m, with preponderance for higher l.
  - P_m increases linearly with a reduction in r due to stress field modifications in the adhesive.
- **Dual adhesives**: The T-joints mostly failed between the two adhesives in the bondline, despite variations with t. The dual-adhesive joint (1/3) provides the best results, and it is possible to obtain relevant P_m improvements over single-adhesive joints.
- **Composite adherends**: A higher l increases P_m up to 20 mm. For a higher l leads to a failure mode modification that reduces the joint performance.

References

1. Petrie EM. *Handbook of Adhesives and Sealants.* McGraw-Hill, New York, USA, 2000.
2. Da Silva LF, Öchsner A, and Adams RD. *Handbook of Adhesion Technology.* Springer Science & Business Media, 2011.
3. Adams RD. *Adhesive Bonding: Science, Technology and Applications.* Elsevier, 2005.
4. Li W, Blunt L, and Stout KJ. Analysis and design of adhesive-bonded tee joints. *International Journal of Adhesion and Adhesives,* 1997; **17**: 303–311.
5. Oliveira JJG, Campilho RDSG, Silva FJG, *et al.* Adhesive thickness effects on the mixed-mode fracture toughness of bonded joints. *Journal of Adhesion,* 2019; 1–21.
6. Luo H, Yan Y, Zhang T, *et al.* Progressive failure and experimental study of adhesively bonded composite single-lap joints subjected to axial tensile loads. *Journal of Adhesion Science and Technology,* 2016; **30**: 894–914.
7. Moya-Sanz EM, Ivañez I, and Garcia-Castillo SK. Effect of the geometry in the strength of single-lap adhesive joints of composite laminates under uniaxial tensile load. *International Journal of Adhesion and Adhesives,* 2017; **72**: 23–29.
8. Ferreira CL, Campilho R, and Moreira RDF. Experimental and numerical analysis of dual-adhesive stepped-lap aluminum joints. *Proceedings of the Institution of Mechanical Engineers, Part E: Journal of Process Mechanical Engineering,* 2020; **234**: 454–464.

9. Raphael C. Variable-adhesive bonded joints (stresses in ordinary lap joint compared to variable adhesive joint), in *Structural Adhesives Bonding, Symposium,* Stevens Institute of Technology, Hoboken, NJ, 1966, pp. 99–108.
10. Hart-Smith LJ. *Adhesive-Bonded Single-Lap Joints,* 1973.
11. Da Silva LF and Campilho RD. Advances in numerical modelling of adhesive joints, in *Advances in Numerical Modeling of Adhesive Joints.* Springer, 2012, pp. 1–93.
12. Belytschko T and Black T. Elastic crack growth in finite elements with minimal remeshing. *International Journal for Numerical Methods in Engineering,* 1999; **45**: 601–620.
13. Zhan X, Gu C, Wu H, *et al.* Experimental and numerical analysis on the strength of 2060 Al–Li alloy adhesively bonded T joints. *International Journal of Adhesion and Adhesives,* 2016; **65**: 79–87.
14. Barzegar M, Moallem MD, and Mokhtari M. Progressive damage analysis of an adhesively bonded composite *T*-joint under bending, considering micro-scale effects of fiber volume fraction of adherends. *Composite Structures,* 2021; **258**: 113374.
15. Ramezani F, Ayatollahi MR, Akhavan-Safar A, *et al.* A comprehensive experimental study on bi-adhesive single lap joints using DIC technique. *International Journal of Adhesion and Adhesives,* 2020; **102**: 102674.
16. Moreira RDF and Campilho RDSG. Strength improvement of adhesively-bonded scarf repairs in aluminium structures with external reinforcements. *Engineering Structures,* 2015; **101**: 99–110.
17. Ribeiro F, Campilho R, Carbas R, *et al.* Strength and damage growth in composite bonded joints with defects. *Composites Part B: Engineering,* 2016; **100**: 91–100.
18. Campilho RDSG, Banea MD, Pinto AMG, *et al.* Strength prediction of single- and double-lap joints by standard and extended finite element modelling. *International Journal of Adhesion and Adhesives,* 2011; **31**: 363–372.
19. Campilho RD, Banea MD, Neto J, *et al.* Modelling adhesive joints with cohesive zone models: Effect of the cohesive law shape of the adhesive layer. *International Journal of Adhesion and Adhesives,* 2013; **44**: 48–56.
20. Faneco T, Campilho R, Silva F, *et al.* Strength and fracture characterization of a novel polyurethane adhesive for the automotive industry. *Journal of Testing and Evaluation,* 2017; **45**: 398–407.

21. Campilho RD, Banea MD, Pinto AM, *et al.* Strength prediction of single- and double-lap joints by standard and extended finite element modelling. *International Journal of Adhesion and Adhesives,* 2011; **31**: 363–372.

22. Campilho RDSG, Banea MD, Neto JABP, *et al.* Modelling of single-lap joints using cohesive zone models: Effect of the cohesive parameters on the output of the simulations. *Journal of Adhesion,* 2012; **88**: 513–533.

23. Campilho RDSG, Banea MD, Chaves FJP, *et al.* Extended finite element method for fracture characterization of adhesive joints in pure mode I. *Computational Materials Science,* 2011; **50**: 1543–1549.

24. Carneiro MAS and Campilho RDSG. Analysis of adhesively-bonded *T*-joints by experimentation and cohesive zone models. *Journal of Adhesion Science and Technology,* 2017; **31**: 1998–2014.

Chapter 8

High-Performance Concrete Made of Seawater and Cured in Seawater to Evaluate Hydration Composition and Porosity with Analysis of SEM, Petrography, and Mercury Porosimetry in the Lack of Fresh Water of Cleaner Built Environment

Erniati Bachtiar,[a] Fatmawaty Rachim,[a] Ritnawati Makbul,[b] Arbain Tata,[c] Muhammad Syarif,[d] Mehmet Serkan Kirgiz,[e] André Gustavo de Sousa Galdino,[f] Konstantinos G. Kolovos,[g] and Andi Yusri[d]

[a]*Department of Civil Engineering, Fajar University, Makassar, South Sulawesi, Indonesia*
[b]*Master in Infrastructure and Environmental Engineering Faculty of Engineering, Fajar University, Makassar, Indonesia*
[c]*Department of Civil Engineering, Universitas Khairun, Ternate, Indonesia*
[d]*Engineering Faculty, Univeristas Muhammadiyah Makassar, Indonesia*
[e]*Academia of ACMCEN, Beaverton, OR 97075, USA*
[f]*Federal Institute of Education, Sciences and Technology of Espírito Santo, Av. Vitória, 1729 Jucutuquara, Vitória, ES, 29040-780, Brazil*
[g]*Hellenic Military Academy, Laboratory of Inorganic Chemistry, Vari, ATTICA, Greece*
nakres42@yahoo.com

Handbook of Innovative Adhesive Technology: Biodegradable and Cost-Effective Binder for Built Environment and Engineering Sciences
Edited by Mehmet Serkan Kirgiz

ISBN 978-981-5129-51-9 (Hardcover), 978-1-003-59892-3 (eBook)
www.jennystanford.com

The marine environment is an aggressive environment that can affect the materials used, especially reinforced concrete buildings. In making supporting infrastructure, the chance of materials getting contaminated with seawater is sometimes unavoidable. This chapter aims to determine the mechanical (compressive and split-tensile strength) and microstructural (product hydration, pore type and size, absorption, and porosity) properties of high-strength concrete (HSC) treated or submerged in seawater. The HSC specimen was made using a water–cement ratio of 1.1, which obtained a water–cement factor of 0.35. The specimen was treated or submerged for 1, 3, 7, 28, and 90 days, according to the age of the test. The test of porosity characteristics and microstructure properties was carried out as per the ASTM standards and international scientific rules. The results showed that concrete's primary product hydration (calcium silicate hydrate [C–S–H] and calcium hydroxide [CH]) continued to increase with the submersion time. Friedel's salt and ettringite were formed on the concrete submerged in seawater for 1–90 days due to chlorides and sulfates in the cement bonding with seawater.

8.1 Introduction

Earth's surface has an extensive sea area. The sea contains a huge quantity of various natural resources that can meet human needs in the world. These needs can be seen in food sources (fish and marine plants), energy sources (petroleum and wave powerplants), transportation, and tourist attractions. Therefore, humans can take advantage of it as much as possible. For utilizing these potentials, various supporting infrastructures have been built on the seacoast or above/in the sea, which include seaports, offshore platforms, bridges, resting places (rest area), etc. The marine area has an aggressive environment. The construction of infrastructure in an aggressive environment such as the sea needs to pay attention to the materials used. Buildings in an aggressive environmental area must be designed by considering the impact of the environment on the structure and materials. The sea area is an aggressive environment that can affect the materials used, especially concrete and steel buildings. Infrastructure construction is generally made of concrete and steel. The concrete used must be of high quality if used in an aggressive environment. As we know, the two materials, if contaminated with seawater, will affect the quality of construction. Seawater containing salt solutions, especially sodium chloride (NaCl) and magnesium sulfate ($MgSO_4$), leads to the corrosion of concrete and steel [1, 2]. The presence of sulfate salts harms the hardened concrete matrix due to the increasing amount of ettringite formation, resulting in a significant increase in volume in the hardened concrete matrix. The large volume in the matrix creates internal stress that cracks the concrete mass and affects its resistance [1].

On the other hand, concrete materials used in the construction of buildings in aggressive areas, in this case at sea, need high-quality concrete. HSC has high strength and low porosity and can inhibit or minimize the entry of seawater to the concrete reinforcement. Besides, HSC has high durability to avoid or minimize the corrosion process of concrete and reinforcement. Previous research on high-performance concrete, namely ultra-high concrete, using seawater and sea sand, reached a compressive strength of up to 180 MPa [3]. In building supporting infrastructure in coastal areas, contact between the materials used and seawater is sometimes unavoidable. Before the concrete hardens or reaches its maximum strength at the age of 28 days, it may be contaminated with seawater; thus, the chemical reactions between cement, water, and seawater cannot be avoided. During the hydration process, the concrete continues to run from 1 day to the age of 28 days. If the concrete-curing process is good, the hydration process will be optimal so that the expected quality of the concrete will be appropriate. However, suppose the concrete is not properly cared for, the hydration process will not be optimal. With increasing age, there will be a decrease in strength and increased porosity and absorption in the concrete [4]. Previous research on HSC that used seawater as mixing and treatment water had higher initial strength than the concrete that used freshwater as mixing water [5, 6]. In addition, research on mortar and concrete using seawater showed that the strength of mortar and concrete increased when using seawater as mixing water [7]. Research on high-performance concrete using seawater and sea sand as mixing water and curing caused a slight decrease in the workability, density, and modulus of elasticity. Besides, it also tended to increase the initial strength slightly, yet slightly reduce it at 7 days and over [3].

Previous studies have analyzed the strength of concrete using seawater as mixing and treatment water, and the effect of seawater on the strength and durability of concrete. However, this study benefits constructions in marine areas that utilize high-performance concrete materials using freshwater as mixing water. However, making concrete cannot avoid seawater contamination; therefore, it will make the concrete to undergo hydration until it hardens. This chapter describes the effects of seawater on the mechanical (compressive and tensile strengths) and microstructural (hydration and porosity processes) properties of HSC. The mechanical and microstructural properties were reviewed at the age of 1 day to 90 days. The durability, as well as mechanical and microstructural properties, of high-performance concrete exposed to seawater needs to be examined until the age of 5 years to analyze its durability prediction on the corrosion of concrete and steel.

8.2 Materials and Method

8.2.1 Materials

The high-quality concrete used local materials, namely, freshwater, aggregate (crushed stones and sand), binder (Portland composite cement), and superplasticizer. The physical properties of the concrete constituents can be seen in Table 8.1. In addition, this study used seawater as an engineering material if the concrete made in an aggressive environment (at sea) was stored in a water bath measuring 2 m × 1.5 m × 0.5 m.

Table 8.1 Physical properties of concrete elements

Material	Type/Properties
Crushed stone	Specific gravity (saturated surface dry): 2.62 Fineness modulus: 6.73 Water content: 1.67% Water absorption: 2.31% Nominal maximum size: 19 mm

Material	Type/Properties
Sand	Specific gravity (saturated surface dry): 2.57 Fineness modulus: 2.54 Water content: 2.94% Water absorption: 1.37% Nominal maximum size: 4.75 mm
Binder	Portland composite cement
Water	Normal tap water
Superplasticizer	Sika-Viscocrete-3115 ID Specific gravity 1.04 ± 0.01 kg/L
Seawater	Salinity 29.4 ppm (part per million)

8.2.2 Research Methods

This study utilized experimental techniques. The Civil Engineering Laboratory of Fajar University manufactured and tested samples of HSC to determine their mechanical characteristics. This study was divided into many stages: material preparation, mixture design, specimen manufacture, curing/immersion, and mechanical and microstructural properties test. The experimental procedure is shown in Fig. 8.1.

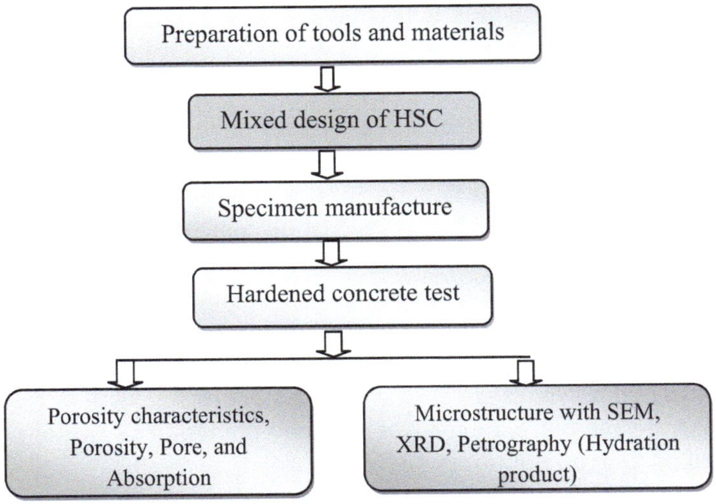

Figure 8.1 Flowchart of the research experiment.

This high-performance mixed-design concrete used the method of the European Guidelines for Self-Compacting Concrete [8]. The mixed-design HSC used a water-to-cement ratio of 1.1 (EFNARC: 0.85–1.1). The composition of the concrete mix with the water–cement factor was obtained at 0.35. Table 8.2 shows the composition of high-strength concrete ingredients (kg/m^3).

Table 8.2 Composition of high-strength concrete ingredients (kg/m^3)

Water	Cement	Sand	Crushed Stone	Superplasticizer
204.01	601.36	770.12	785.11	6.01

The concrete mixing process used a mixer. After all the mixture compositions were homogeneous, the mixture was poured into a cylindrical mold (in a new concrete condition). After the concrete hardened, the specimen was removed from the mold and placed in a water bath filled with seawater. The specimen consisted of three samples for each test variation for compressive and split-tensile strengths and the age of curing/immersion of the specimen. Treatment or immersion of the concrete samples was carried out in a water bath filled with seawater. The treatment times or immersion of specimen using seawater were 1, 3, 7, 28, and 90 days. Mechanical and microstructural properties were tested after the concrete samples were 1, 3, 7, 28, and 90 days old. The pore type and size microstructure used a petrography tool based on the ASTM C295 standard [9], and the porosity test of concrete was determined using the ASTM C642-13 standard [10].

The microstructural analysis used scanning electron microscopy (SEM), X-ray powder diffraction (XRD), and petrography tests. The XRD analysis was done with the Rigaku Miniflex II XRD tool, the SEM analysis with the vega3tescan SEM-EDX tool, and the petrography test with an Olympus BX 51-P polarizing microscope. The SEM test was used to analyze the morphology of the concrete microstructure. The XRD test analyzed the quantity of the microstructural phases (compounds) in the concrete, namely, tobermorite, portlandite, ettringite, Friedel's salt, alite, belite, aluminate, and ferrite. In contrast, the petrographic test analyzed the type and size of pores and porosity in HSC concrete.

The ASTM standard C642-13 was used to determine porosity: The specimen's mass was determined, and the specimen was dried for at least 24 h at 100–110°C. The specimen was submerged in water at about 21°C for at least 48 h, and up to two consecutive grades of dry sample surface mass at 24 h intervals demonstrated a rise in the mass of less than 0.5% of the bigger value. Then, in an appropriate container, it was simmered for 5 h and cooled naturally for at least 14 h or until the final temperature reached 20–25°C. The mass of the specimen was determined after it was cooked and then again after it was cooled (with wire, and the apparent mass in water was determined). Equation 8.1 might be used to determine the porosity:

$$e = \frac{Y - X}{Y - Z} \tag{8.1}$$

where e denotes the porosity of the concrete, X denotes the mass of an oven-dried sample in the air, Y denotes the bulk of a surface-dry sample in the air after immersion and boiling (g), and Z denotes the mass of a sample in water following immersion and boiling (g).

8.3 Results and Discussions

8.3.1 Hydration Products and Changing in Friedel's Salt ($3CaO{\cdot}Al_2O_3{\cdot}CaCl_2{\cdot}10H_2O$) and Chlorine ($Cl_2$) with XRD

The microstructure is the type, amount, size, shape, and distribution of the phases present in a solid. The microstructure is the extremely small-scale structure of a material, defined as the structure of a material's prepared surface as revealed by an optical microscope at a magnification of more than 25×. The microstructure of a material (concrete, mortar, polymers, metals, ceramics, or composites) may significantly affect physical characteristics such as strength, hardness, ductility, toughness, corrosion resistance, high/low-temperature behavior, and wear resistance. These characteristics, in turn, dictate how these materials are used in industrial practice [11]. The material microstructure will affect the strength or quality of concrete, as previous research [12, 13] so it has much territory that the quality of the source water is not qualified as mixing water in construction. Besides, construction of concrete in areas that are likely quantity of water or fresh water is very minimal or even nothing then the sea water cannot be avoided in mixing concrete. This research was an experimental study, the samples for compressive strength test are cylindrical premises size of 10 mm × 20 mm. The porosity relation, compressive power, age and model of porosity relationship with SCC concrete compressive strength which using sea water is discussed in this paper. Compressive strength testing is following the standard ASTM 39/C 39-99. Universal Testing Machine (UTM about concrete using seawater as its mixing water showed that the smaller the concrete porosity, the better the concrete quality or strength.

Theoretically, the hydration heat growth rate is related to the concrete age. As the age increases, the hydration heat increases too, and this process continues until the age of 90 days, reaching 96% hydration process [14, 15]. The main products of the hydration process between the chemical reaction of cement and water are C–S–H and CH (portlandite), as described in formulas 8.4 and 8.5 [16].

$$C_3S + H_2O \longrightarrow C\text{–}S\text{–}H + Ca(OH)_2 \quad (8.2)$$

$$C_2S + H_2O \longrightarrow C\text{–}S\text{–}H + Ca(OH)_2 \quad (8.3)$$

The main hydrates of cement paste reacting with water are tobermorite (C–S–H gel), portlandite ($Ca(OH)_2$), ettringite/AFt ($C_3A{\cdot}3CaSO_4{\cdot}32H_2O$), and AFm ($C_3A{\cdot}3CaSO_4{\cdot}10H_2O$) [14, 16]. Of these hydrates, AFt (aluminoferrite tetrasulfate) and $Ca(OH)_2$ have a small capacity to bind chlorides. C–S–H has a very large surface area and can bind a wide range of surfaces and a variety of ions, including chloride ions. Furthermore, the chloride-binding capacity of C–S–H depends on the chemical composition and surface area, the type of chloride solution, and the experimental conditions. Based on the two hydration phases that can bind more chloride (C–S–H and AFm [aluminoferrite monosulfate]), there are two main chloride-binding mechanisms: physical adsorption and chemical reactions. The C–S–H phase binds chloride through absorption. In contrast, AFm binds chloride via chemical reaction to form Friedel's salt crystals [14–16].

Previous studies showed that aggressive environments, such as the sea, contained sulfates or chlorides; cementitious materials could bind sulfates and chlorides to form an ettringite (AFt, $3CaO{\cdot}Al_2O_3{\cdot}3CaSO_4{\cdot}32H_2O$) and Friedel's salt ($3CaO{\cdot}Al_2O_3{\cdot}CaCl_2{\cdot}10H_2O$) [16–18]. The formation process can be seen in formulas 8.6 and 8.7 [18].

$$3CaO + Al_2O_3 + 3CaSO_4 + 32H_2O \rightarrow 3CaO{\cdot}Al_2O_3{\cdot}3CaSO_4{\cdot}32H_2O \quad (8.4)$$

$$3CaO + Al_2O_3 + CaCl_2 + 10H_2O \rightarrow 3CaO{\cdot}Al_2O_3{\cdot}CaCl_2{\cdot}10H_2O \quad (8.5)$$

The results of the microstructural test on high-performance concrete using an XRD device can be seen in Table 8.3. The highest hydration process products were C–S–H and CH gels. As previous researchers stated, the two main hydrate products from cement paste reacting with water were tobermorite (C–S–H gel) and portlandite ($Ca(OH)_2$) [14, 16]. In addition, the influence of seawater containing sulfate and chloride occurred on HSC bonds, which formed Friedel's salt and ettringite. Friedel's salt was formed from 7 to 90 days, which increased with age, while ettringite was formed from 1 to 90 days. This ettringite was formed from the start because of the AFt and AFm content in the cement. Thus, the possibility of ettringite formation at the beginning of the concrete age was due to the content of AFt and AFm in the cement. This ettringite, at the age of 28 to 90 days, continued to increase, which could cause concrete corrosion (concrete swelling) and peeling, so that an aggressive environment could reach concrete reinforcement. Table 8.3 shows that the concrete aged with treatment or seawater submersion.

Table 8.3 Hydration products of HSC

Microstructure Phase (Hydration Products)	**Age (Days)/Percentage (%)**				
	1	**3**	**7**	**28**	**90**
Portlandite, $Ca(OH)_2$	12.31	13.74	14.81	19.57	20.57
Tobermorite, $3CaO{\cdot}2SiO_2{\cdot}3H_2O$	20.51	25.31	40.21	42.50	45.94
Friedel's salt, $3CaO{\cdot}Al_2O_3{\cdot}CaCl_2{\cdot}10H_2O$	0.00	0.00	1.59	5.00	6.40
Ettringite, $3CaO{\cdot}Al_2O_3{\cdot}CaSO_4{\cdot}32H_2O$	0.21	1.34	2.12	3.26	7.66
Chlorine, Cl_2	0.00	0.00	1.06	2.17	3.77
Tricalcium silicate, $3CaO{\cdot}SiO_2$	28.72	23.97	13.76	7.93	5.37
Dicalcium silicate, $2CaO{\cdot}SiO_2$	38.26	35.64	26.46	19.57	10.29
Tricalcium aluminate, $3CaO{\cdot}Al_2O_3$	—	—	—	0.00	0.00
Tetracalcium aluminoferrite, $4CaO{\cdot}Al_2O_3{\cdot}Fe_2O_3$	—	—	—	0.00	0.00
Weight (%)	100	100	100	100	100

This table especially shows the number of hydration products remained and increased, such as tobermorite (C–S–H) produced from the reaction of tricalcium silicate (C_3S) and dicalcium silicate (C_2S) with water (H_2O) and seawater. The microstructure of the hydration product formed in the concrete decreased the content of dicalcium silicate ($2CaOSiO_2$) and tricalcium silicate ($2CaOSiO_2$) due to reacting with water to form C–S–H gel or tobermorite ($3CaO{\cdot}2SiO_2{\cdot}3H_2O$) and portlandite ($Ca(OH)_2$), resulting in the presence of increased content of tobermorite ($3CaO{\cdot}2SiO_2{\cdot}3H_2O$). The microstructure formed in the concrete contaminated with seawater decreased the content of dicalcium silicate ($2CaOSiO_2$) and tricalcium silicate ($2CaOSiO_2$) due to reacting with water to form C–S–H gel or tobermorite ($3CaO{\cdot}2SiO_2{\cdot}3H_2O$) and portlandite ($Ca(OH)_2$). The content of tobermorite ($3CaO{\cdot}2SiO_2{\cdot}3H_2O$) and portlandite ($Ca(OH)_2$) increased along with the age of the concrete up to 90 days. The treatment of specimens in seawater caused the presence of chlorine, and a Friedel's salt phase was formed at the age of 7, 28, and 90 days; this can be seen in Table 8.3 and Figs. 8.2 and 8.3.

8.3.2 Microstructure and Morphology of HSC with SEM

Figure 8.2 shows an image of the HSC microstructure submerged in seawater using an SEM tool with 1000× magnification, with variations in ages of 3 days (a), 7 days (b), 28 days (c), and 90 days (d). The four images appeared to have different microstructures. At the age of 3 days, the presence of salt crystals was not visible. The white salt crystals became clearer at 7, 28, and 90 days. More details can be found in Figs. 8.2 and 8.3. The effect of seawater on the concrete formed salt crystals, as shown in Fig. 8.3. Figure 8.2 illustrates that at the age of 28 days, the concrete was submerged or treated with seawater, resulting in the exposure of ettringite, Friedel's salt, and the principal products of cement paste (CH and C–S–H).

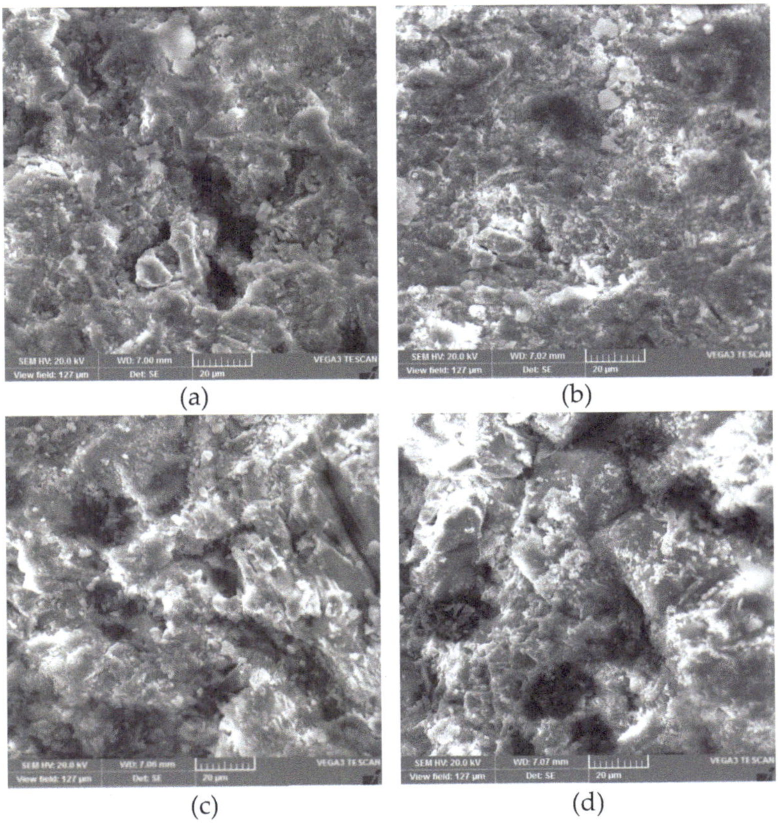

Figure 8.2 Morphology of HSC microstructure using SEM, submerged in seawater for 28 days.

Figure 8.3 shows the morphology of HSC concrete with a magnification of 5000× and 10000×. The XRD results showed that since day 1, the HSC microstructure had formed C–S–H, CH, and ettringite compounds. The difference from the concrete treated with freshwater is that there was no Friedel's salt formed. The XRD results are in Table 8.3, and the morphology can be seen in Fig. 8.5; a white hexagonal shape can be seen in Friedel's salt. In addition, the concrete morphology was also depicted as ettringite shaped like a needle. Ettringite was formed on the concrete from day 1 according to the results of the XRD analysis. Previous research on concrete using seawater as mixing water showed that Friedel's salt and ettringite were formed from 1 to 28 days [17]. The portlandite ($Ca(OH)_2$), produced from the reaction of Portland cement and water, then bonded with NaCl to form calcium chloride, which then reacted with aluminate in cement to form Friedel's salt ($3CaO{\cdot}Al2O_3{\cdot}CaCl_2{\cdot}10H_2O$) [19].

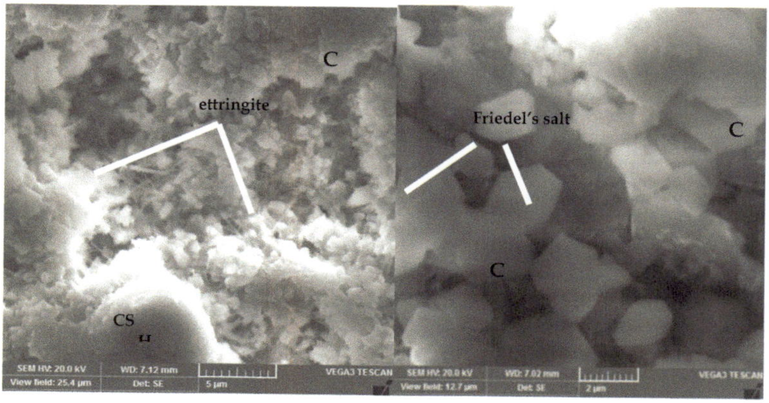

Figure 8.3 Morphology of HSC microstructure using SEM, submerged in seawater for 28 days.

8.3.3 Porosity of Concrete

There were two types of pores observed in the concrete, and they were intergranular and isolated pores. Intergranular pores exist between the grains making up the concrete, more minor than isolated pores. Isolated pores are formed from the beginning of the formation of the sample. Observations of micrograph photos and types and sizes of pores through petrography can be seen in Table 8.4 and Fig. 8.4.

Table 8.4 Results of petrography testing on HSC in seawater

Age (Day)	Pore (%) Concrete Matrix	Aggregate	Amount	Intergranular Pore (μm) min	max	Isolated Pore (μm) min	max
3	14.4	2	16.14	172.7	1106.8	9.2	20.4
7	11.8	2.05	13.54	117.4	346.9	8.3	19.3
28	10.64	2.2	12.38	110.5	249.9	3.3	17
90	10.62	2.05	12.36	89.5	237.7	3.1	13.7

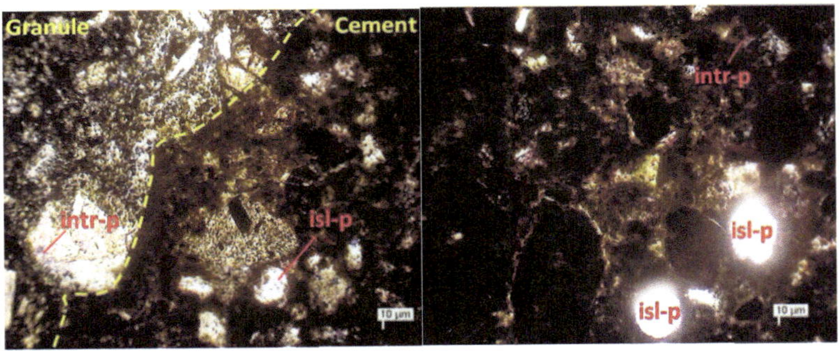

Figure 8.4 Petrography photos of high-strength concrete at the age of 28 days. Isl-p: isolated pores; Intr-p: intergranular pores.

Table 8.4 shows that HSC concrete submerged in seawater for 3 to 90 days showed that the porosity value was getting smaller with age. The pore value in the concrete matrix seemed to decrease with age. However, specifically for the aggregate pore value, the average pore value was 2–2.2%. The intergranular pores looked smaller with age, from 1106.8 μm (3-day old) to 237 μm (90-day old). The isolated pores looked smaller with age, from 20.4 μm (3-day old) to 13.7 μm (90-day old). The isolated and intergranular pores can be seen in Fig. 8.4. Figure 8.4 shows a micrograph photo of 28-day-old concrete using a petrography tool. The concrete had a good matrix bond with the aggregate, with no cracks or channels visible on the micrograph photo. The concrete matrix had isolated pores of 3.3–17.0 μm and intergranular pores of 110.5–249.9 μm.

8.3.4 Absorption of Concrete

The results of the absorption and porosity tests of HSC concrete can be seen in Table 8.5. The observations and calculations of the absorption and porosity of HSC concrete were carried out at the ages of 3, 7, 28, and 90 days.

Table 8.5 Result of absorption and porosity based on ASTM

Age (Day)	Absorption (%)		Porosity (%)	
3	9.93	9.73	12.75	12.97
	9.46		12.67	
	9.80		13.49	
7	8.79	8.57	11.90	12.09
	8.33		11.80	
	8.60		12.57	
28	7.19	6.95	10.66	10.82
	6.73		10.53	
	6.93		11.26	
90	6.20	5.95	10.56	10.71
	5.75		10.43	
	5.90		11.15	

Once the water absorption of concrete was evaluated in terms of porosity of concrete subjected to seawater, it was understood that the absorption and porosity values of HSC concrete decreased with the increasing age of the concrete. To obtain useful information regarding the porosity and water absorption decreasing, it is evident that Friedel's salt expands in voids in HSC subjected to the curing of seawater. Sui *et al.* evaluated the effect of Friedel's salt on the properties of tricalcium silicate-based binder with a combination of metakaolin and limestone. The work unveiled that carbonate participates in Friedel's salt formation and results in the overestimation of Friedel's salt content [20]. The finding of this research is in line with the result of the current paper, which presented a linear decrease in the absorption and porosity of concrete subjected to seawater with time.

8.4 Conclusion

The mechanical properties of high-performance concrete treated or submerged in seawater from 1 to 90 days increase as the concrete ages. Meanwhile, the microstructural properties continue to process as long as the hydration process is ongoing. The primary product hydration (C–S–H and CH) continues to increase with curing time or submersion in seawater. Concrete submerged in seawater at the age of 1–90 days forms Friedel's salt and ettringite due to the bonding of cement with seawater containing chlorides and sulfates. Friedel's salt is formed from 7 to 90 days, which increases with age, while ettringite is formed from 1 to 90 days. The pore value in the concrete matrix seems to decrease with age. However, specifically for the aggregate pore value, the average pore value is 2–2.2%. The intergranular pores look smaller as they aged, from 1106.8 μm (3-day old) to 237 μm (90-day old). The isolated pores look smaller as they aged, from 20.4 μm (3-day old) to 13.7 μm (90-day old).

References

1. Qasim OA, Maula BH, Moula HH, and Jassam SH. Effect of salinity on concrete properties. *IOP Conf Ser Mater Sci Eng.* 2020; 745. https://doi.org/10.1088/1757-899X/745/1/012171.
2. Mather B and Engineer C. Effects of seawater on iron. *J Franklin Inst.* 1837; **23**: 483. https://doi.org/10.1016/s0016-0032(37)91392-9.
3. Teng J-G, Xiang Y, Yu T, and Fang Z. Development and mechanical behaviour of ultra-high-performance seawater sea-sand concrete. *Adv Struct Eng.* 2019; **22**: 3100–3120. https://doi.org/10.1177/1369433219858291.
4. Erniati B. The self-compacting concrete (SCC) using seawater as mixing water without curing. *ARPN J Eng Appl Sci.* 2018; **13**: 4057–4061. https://doi.org/10.31227/osf.io/pujv7.
5. Erniati, Tjaronge MW, Djamaluddin R, and Sampebulu V. Porosity and microstructure phase of self-compacting concrete using sea water as mixing water and curing. *Adv Mater Res.* 2015; **1119**: 647–651. https://doi.org/10.4028/www.scientific.net/amr.1119.647.
6. Erniati, Tjaronge MW, Djamaluddin R, and Sampebulu V. Compressive strength and slump flow of self-compacting concrete uses fresh water and sea water. *ARPN J Eng Appl Sci.* 2015; **10**: 2373–2377.

7. Irmawaty R and Tjaronge MW. Effect of seawater as mixing water on the mechanical properties of mortar and concrete. *ConCERN (Conference Civ. Eng. Res. Networks)* 2014; **2014:** 61–64.
8. EFNARC. *The European Guidelines for Self-Compacting Concrete: Specification, Production and Use.* UK, 2005.
9. ASTM C295. *Standard Guide for Petrographic Examination of Aggregates for Concrete* 2019.
10. ASTM C642-13. *Standard Test Method for Density, Absorption, and Voids in Hardened Concrete*, West Conshohocken, PA, 2013.
11. Bimbo AD, Farinella GM, Escalante HJ, Bertini M, Cucchiara R, Vezzani R, *et al. Pattern Recognition. ICPR International Workshops and Challenges*, vol. 12664, Springer International Publishing, Cham, 2021. https://doi.org/10.1007/978-3-030-68799-1.
12. Erniati, Tjaronge MW, Zulharnah, and Irfan UR. Porosity, pore size and compressive strength of self-compacting concrete using sea water. *Procedia Eng.* 2015; **125**: 832–837. https://doi.org/10.1016/j.proeng.2015.11.045.
13. Erniati, Tjaronge MW, Djamaluddin R, and Sampebulu V. Microstructure characteristics of self-compacting concrete using sea water. *Int J Appl Eng Res.* 2014; **9**: 18087–18095. https://doi.org/10.31224/osf.io/6jnfv.
14. Chen M, Geng J, Xiong H, Shang T, Xue C, and Abbas M. Effect of curing on mechanical properties of cement-stabilized coral sand in marine environment. *Adv Mater Sci Eng.* 2020; **2020**. https://doi.org/10.1155/2020/4678376.
15. Bai-zhou P and Zhuo-bin W. Experimental study on effects to coral sand concrete compressive strength of raw materials. *Eng Mech.* 2015; **32**: 221–225.
16. Lin Y, Xu D, and Zhao X. Properties and hydration mechanism of soda residue-activated ground granulated blast furnace slag cementitious materials. *Materials (Basel)* 2021; **14**: 1–16. https://doi.org/10.3151/coj.52.387.
17. Mansyur DP. Mechanical behaviour and microstructure characteristic of concrete by using freshwater and seawater. *Civ Eng J.* 2020; **6**: 1195–1203. https://doi.org/10.28991/cej-2020-03091540.
18. Maes M and De Belie N. Influence of chlorides on magnesium sulphate attack for mortars with Portland cement and slag based binders. *Constr Build Mater.* 2017; **155**: 630–642. https://doi.org/10.1016/j.conbuildmat.2017.07.201.

19. Halim LN, Ekaputri JJ, and Triwulan. The influence of salt water on chloride penetration in geopolymer concrete. *MATEC Web Conf.* 2017; **97**. https://doi.org/10.1051/matecconf/20179701002.

20. Sui S, Wu M, Yang Z, Wang F, Liu Z, and Jiang J. An investigation on the formation of Friedel's salt in tricalcium silicate combined with metakaolin and limestone systems. *Constr Build Mater.* 2021; **284**. https://doi.org/10.1016/j.conbuildmat.2021.122855.

Chapter 9

Portland Cement and Starch-Based Adhesives: A Comprehensive Review

Konstantinos G. Kolovos[a] and Mehmet Serkan Kirgiz[b]
[a]*Hellenic Military Academy, Laboratory of Chemical Technology, 16673 Vari, Attica, Greece*
[b]*Academia of ACMCEN, Beaverton, OR 97075, USA*
nakres42@yahoo.com

In the last decades, there has been a growing interest in the development of environmentally friendly adhesive materials for green concrete, as an important step to mitigate the intense environmental impacts of construction activities. Starches are renewable, abundantly available polymers, which can potentially be considered a feasible option for the synthesis of bio-admixtures that can be efficiently used in sustainable green construction. The use of different sources of starch as chemical admixtures and their adhesive behavior and influence on Portland cement-based composites and concrete properties are summarized. The progress in research on natural or synthetic starch admixtures regarding their effect on Portland hydration process, their potential use as temperature

Handbook of Innovative Adhesive Technology: Biodegradable and Cost-Effective Binder for Built Environment and Engineering Sciences
Edited by Mehmet Serkan Kirgiz

ISBN 978-981-5129-51-9 (Hardcover), 978-1-003-59892-3 (eBook)
www.jennystanford.com

rise inhibitor (TRI) materials or as water-reducing agents for the improvement in the rheological properties of fresh concrete or materials for 3D printing, and their influence on the mechanical properties of different types of hardened concrete materials is also discussed.

9.1 Introduction

It is widely known that to the present, the building and construction sector is the most chemical-intensive sector downstream of the chemical industry. It is the largest end market for chemicals and the sector generating the highest chemical revenue [1]. It is estimated that the cement industry is responsible for more than 7% of the global CO_2 gas emissions, a fact that urges for the development and implementation of new efficient technologies toward the heavily emitting production process of Portland cement, which is the basic binder material used to produce concrete—one of the fundamental construction materials of our times. The global production of Portland cement has reached 10 billion metric tons per year and is ranked second among the largest CO_2-emitting industries in the world, being the third largest energy-demanding material globally [2].

The above trend activates the need for renewable resources as materials for structural applications, which have been gaining interest. In the search for alternative materials in structural applications, starch and cellulose have become potential alternative candidates for selected purposes due to a number of desirable properties that involve nontoxicity, degradability, low cost, and wide availability [3–5].

9.2 Starch Chemistry and Properties

Starch (or amylum) is a natural polysaccharide polymer produced by all green plants in the form of a white, tasteless powder. Its solubility in water is negligible as well as in other common organic solvents such as alcohol. The proposed basic chemical formula of the starch molecule is $(C_6H_{10}O_5)_n$+(H_2O) where the glucose monomers are bonded in α-1,4 linkages. Two forms of pure starch can be clearly distinguished: the first involves a linear and helical polymer called amylose, and the second refers to the branched form named amylopectin [6]. The green leaves of plants can produce starch by the excessive glucose molecules during photosynthesis. The produced starch is then stored to serve as a reserve food supply in chloroplasts, in the form of granules or in other storage organs such as roots of cassava plants, tubers of potatoes, stem pith of sago, and seeds of corn, wheat, and rice.

It is widely proven that starch is a cheap, widely available, and easily biodegradable material that has been extensively used in food, medicine, and agricultural products [7]. Amylose and amylopectin are present in various ratios depending basically on the botanical source. Sources of starch involve fruit and vegetable wastes, wastes produced by the food industry as well as wastes derived by paper and fermentation industries [7, 8].

Due to the various sources of starch, properties vary in terms of the chemistry of the polymer affecting its hydrophobical performance. This is attributed to the starch hydroxyl groups that can easily form hydrogen bonds with water, hence increasing hydrophilicity [9]. Solutions for enhancing hydrophobical properties include the bonding addition of hydrophobic functional groups such as esters onto native starch chains to improve their water resistance through chemical, mechanochemical, or enzymatic processes.

9.3 Starch-Based Adhesives

It has been found that the paste formed when starch is mixed with warm water presents properties suitable for use as a thickening, stiffening, or gluing agent. Starched-based gluing materials were initially made from potato starch glues, as well as other known types such as wheat and maize starch. Hence, one of the major uses of starch is as an adhesive material for different purposes, even though the usage of native starches is limited due to their relatively poor shear strength, low thermal stability, and hydrophilicity [10], which can be overpassed by chemical modification techniques to alter the structure and properties of biopolymers. These techniques involve not only esterification as stated earlier but also etherification, oxidation, crosslinking, and grafted copolymerization. Starch modification by oxidation and subsequent cross-linking can provide adhesives exhibiting higher shear strengths and lower tendency for retrogradation [8, 11]. Starch adhesives are much in dry form, cold soluble, and hot soluble.

9.3.1 Starch and Cement-Based Materials

It has been showed in related studies that native or different forms of waste-recovered starch can be a promising material in applications that involve classic cement-based building materials [12]. Cement-based composites produced with the addition of starch and used as a binder exhibited increased flowability and setting time as well as low thermal conductivity without significant deterioration of mechanical strength properties [7]. Moreover, starch can be used to produce low-cost biodegradable construction materials in the absence of a cement matrix. In a previously published work by Kulshreshtha *et al.* [13], a very interesting biodegradable novel construction material (named CoRncrete) was developed by heating a mixture of corn starch with sand and water, which showed cost-effective and eco-friendly properties and satisfactory compressive strength comparable to that of common red clay bricks. Up to a certain extent, the physical and morphological characteristics of CoRncrete product resemble the properties of a sandstone material that uses gelatinized starch as a binder and sand as a filler, while exhibiting the thermoplastic mechanical behavior of polymer-matrix composite materials.

Mansour *et al.* studied the preparation and properties of a starch–sand composite for industrial applications [14]. The overall conclusion was that these artificial starch-based sandstone materials, after appropriate treatment for increasing hydrophobicity, can behave as polymer-matrix composites with thermoplastic mechanical behavior. Mechanical tests have shown that the compressive strengths measured for the samples were 30 MPa, while the tensile strength to compressive strength ratio reached a value of approximately 0.20. Additionally, significantly low values of modulus of elasticity were found, less than 2 GPa.

As a renewable biological material, native starch has little influence on the hydration kinetics of Portland cement [15, 16], but it is commonly used in the cement industry as the raw material of starch-based admixtures to improve the water retention capacity

[17] and rheological properties [18] and sometimes to retard [15, 16] or control [19] the hydration process.

The final product of starch hydrolysis is glucose, which exerts a well-known retarding effect on cement hydration [20]. In the cement pore solution, glucose promotes calcium complexation and may poison the reaction site and/or the nuclei of cement hydrates, thus retarding cement hydration by adsorbing non-selectively on the inorganic surfaces and disturbing the initial formation, rather than the growth of hydrates [21].

According to Yan and Wang [20], there is a potentially depressing ability of starch-based admixtures to change heat release during the main hydration peak, and the results of the research on the impact of starch-based admixtures on the exothermic process of cement hydration have proven that the exothermic process of cement hydration can be significantly affected by the addition of starch-based admixtures. The action times for sugars and soluble dextrins are restricted by the quick dissolution and adsorption properties of cement paste, leading to a preferential influence on the induction period. Starch-based admixtures affect both the appearance time and maximum heat flow of the main hydration peak. Similar research findings from Yan and Scrivener [22] on a novel starch-based admixture, named temperature rise inhibitor (TRI), specially designed to modify the exothermic process of cement hydration and solve thermal cracking issues, are in agreement with the above. These findings were in line with the published data from other research groups [23, 24] where the retarding effect on the hydration of cement induced by starch TRI dissolved in water and NaOH is controlled by the organic molecular size, and conclusions were verified by testing the effect of the same % wt. additions of dextrans having different molecular sizes on cement hydration.

Afroz *et al.* studied the potential of starch and arrowroot to be used as admixtures as well as their effects on different properties of cement paste and mortar [25, 26] and found that small amounts of starch admixture, ranging from 0.5 to 2.5% wt. of cement content, increased the flowability and setting time of cement mortar, keeping the strength, durability, and shrinkage parameters within an acceptable range, whereas 1.0–1.5% starch addition by weight of cement exhibited improved performance in terms of setting time and flow.

A thorough investigation on the influence of various modified starches on the rheological properties of cement composites and their influence on the mechanical properties of hardened cement mortars has been conducted by Sybis and Konował [27], and the main findings reveal an effect on the fluidization of cement mixes, since a significant decrease in the yield points and plastic viscosities of the cement slurries, attributed to the higher saccharification degree of shorter polymer chains in starch hydrolysates, has also affected the consistency and compressive strength of the cement mortars depending on the type of modification and the length of the starch type polymer chain.

Additional findings were published by Sybis *et al.* [28] on the physicochemical properties of cement composites and concrete where dextrin—a different starch hydrolysate additive—was used, at addition percentages from 0.0 to 0.7% wt. in relation to the cement weight, for samples made with water–cement ratios ranging from 0.40 to 0.50. It was recorded that even the minimum quantity of dextrin addition at a water–cement ratio of 0.40 resulted in 40% increase in compressive strength (up to 67.9 MPa) and improved rheological properties, attributed to the surfactant action of dextrin, possibly related to the short carbohydrate chains, at a given range of concentrations.

9.3.2 Starch in Concrete

Starch has been used in combination with other common mineral admixtures to produce typical and special concretes types with ranging results in the obtained physical and mechanical properties [7, 12, 29–38], rheological behavior [39–41], and durability of the produced syntheses [42, 43].

Cassava starch has been used as a natural admixture in fresh and hardened concrete production to partially replace cement. In studies by Kone *et al.* and Oni *et al.* [36, 37], concrete mixtures with a water–cement ratio of 0.50—produced with the addition of 1.0–2.0% wt. of cassava starch, or blends of cassava starch with rice husk, and cured for 7, 14, 28, 56, and 90 days—exhibited decreased workability, prolonged setting times, and increased compressive strength at all curing days. The above results were attributed to the viscous cassava starch behavior that reduced bleeding and segregation in concrete and to the higher degree of polymerization of cassava starch, increasing binding and thus enhancing compressive strength.

Similar results have also been published [33] concerning increased compressive, split tensile, and flexural strengths observed in concrete samples where 1.5% wt. maize and 2% wt. tapioca were added as starch sources instead of cassava starch.

The effect of starch addition of up to 2.0% wt. by weight of cement, as admixture, on the creep of concrete mixes, monitored for a period of up to 13 months, extracted from cassava and maize, has been examined by Akindahunsi [35]. Enhancement in concrete strength and decrease in deformation with no adverse effect on concrete in the long term were observed, although creep deformations were determined to be lower in concrete specimens with lower starch dosages. Long-term durability characteristics have also been tested by Akindahunsi and Uzoegbo [42] for concretes with 0.5–2.0% wt. percentages of cassava and maize as admixtures. The incorporation of cassava and maize starches into concrete was found to improve the durability properties in terms of increased strength development and decreased oxygen permeability and sorptivity rates.

The environmental benefits of hemp-based composites for building thermal insulation, in place of traditional materials such as rock wool, have been quantitatively assessed in the relevant literature [31, 32]. Different chemical treatments (alkaline treatment and silane coupling agent) of hemp hurds in starch-based composites were proposed in modifying surface properties to enhance the adhesion with the polymer matrix and also for improved mechanical properties. The variation in mechanical resistances such as compressive and flexural strengths in the function of the starch/hemp ratio was analyzed and discussed. The mechanical performance of a starch–hemp concrete agro-composite material made by the binder solution as adhesive, based on starch as the matrix and hemp as fiber, was examined [31]. It has been found that the compressive strength of the starch–hemp concrete depends lightly on the material density, whereas the tensile strength is strongly dependent. However, the Young modulus obtained for the different compositions examined was very low, thus restricting the use of starch–hemp concrete in construction only as a filling material.

Starch has also been reported to be blended with 70/100 paving grade bitumen and modified binder to prepare a stone mastic asphalt with calcium carbonate as a filler. Modified asphalt concrete containing starch performs better than conventional and styrene–butadiene block copolymer–modified mixture, as it exhibits reduced rutting potential and temperature susceptibility [44].

Other potential use for wheat and corn starch has been reported [45] to include the production of lightweight concrete with low sound and thermal conductivity and its low plastic density. The results have shown that as far as the source of starch was concerned, no significant effect on the physical and mechanical properties of the concrete was observed, although the starch-based lightweight concrete had a lower compressive strength than samples made of perlite aggregate.

The use of corn steep liquor as a low-cost replacement growth medium for supporting microbial-induced carbonate precipitation technology at commercial scale and the impact of the growth medium on several key properties of concrete have been investigated by Joshi *et al.* [30]. Tests conducted in bacterial admixed and bacterial spray-treated specimens that contained corn steep liquor showed no adverse effect on the setting characteristics of cement paste and significant improvement in compressive strength and permeation properties. Improved resistance against the ingress of water and aggressive agents was also reported. Similar findings have also been reported [34] for the addition of corn as an admixture in concrete, which involved increased compressive strength, attributed to the increased workability and density for samples that contained 1.0% wt. corn by weight of cement.

9.3.3 Starch as Water-Reducing Agent in Concrete

Numerous studies in the past [18, 29, 46–51] have experimented with starch (mostly in the form of starch sulfonate) to develop a water-reducing agent for the improvement in rheological properties of fresh concrete. The performance of starch as a water-reducing agent has attained less interest in concrete mix designs, since preliminary results were obscured by the lack of explanation of the theoretical mechanism [52]. However, the dispersing mechanism of starch sulfonate as a water-reducing agent for cement has been explained by Zhang *et al.* [52], who suggested improved properties such as reduced flow loss of the fresh cement paste in dispersions with starch sulfonate by steric hindrance, since cement particles are less affected by the rapid kinetic rates of hydration reactions.

However, the above findings partially inhibit starch as an admixture to improve the thixotropic behavior of cement-based materials used to produce 3D-printed elements [53], due to a lower rate of rigidification because of lower rates of formation of hydration products. According to the studies conducted by Moeini *et al.* [53], the elasticity of mixtures that contained a low molecular weight starch-based viscosity-modifying admixture (VMA) was reduced, as their absorption onto the cement particles is hindered through steric hampering and changes occurred in the viscoelastic properties and the static yield stress of the mixtures. A comparable methodology for the fabrication of a 3D-printable material comprising cellulose ether and starch ether has been put forth by Ivaniuk *et al.* [54], founded upon the tenets of Strain Hardening Cementitious Composite (SHCC). This resulted in a prolonged period of maintaining initial workability, higher shape stability, enhanced surface evenness, and high compressive and bending strengths for the printed specimens. These properties were attributed to the improved water retention caused mainly by the cellulose ether addition and less by the starch ether.

9.4 Conclusion

From this review, it can be stated that the use of starch-based adhesives in Portland cement mortars and typical or special concretes may affect hydration and setting, perform as set retarding and flowability modifying admixtures on fresh concrete with profound effects on physical and mechanical properties as well as on the durability parameters of hardened concrete. This review confirms the potential of starch as a renewable and sustainable admixture, an environmentally friendly material that is abundantly available, requires no pretreatment, and does not pose a threat to ecological balance. Thus, it proposes a potential green construction material leading to sustainability.

References

1. Harris, S., Mata, É., Lucena, A. F. P., and Bertoldi, P. Climate mitigation from circular and sharing economy in the buildings sector. *Resources, Conservation and Recycling*, 2023, **188**, 106709.
2. International Energy Agency (IEA). Technology Roadmap—Low-Carbon Transition in the Cement Industry, 2018 (www.wbcsdcement.org). (Accessed April 10, 2022).
3. Ilyas, R. A., Sapuan, S. M., Ishak, M. R., and Zainudin, E. S. Development and characterization of sugar palmnanocrystalline, cellulose reinforced sugar palmstarch bionanocomposites. *Carbohydrate Polymers*, 2018, **202**, 186–202.
4. Sanyang, M. L., Sapuan, S. M., Jawaid, M., Ishak, M. R., and Sahari, J. Effect of sugar palm-derived cellulose reinforcement on the mechanical and water barrier properties of sugar palm starch biocomposite films. *Bioresources*, 2016, **11**, 4134–4145.
5. Du, H. S., Liu, W. M., Zhang, M. L., Si, C. L., Zhang, X. Y., and Li, B. Cellulose nanocrystals and cellulose nanofibrils based hydrogels for biomedical applications. *Carbohydrate Polymers*, 2019, **209**, 130–144.
6. Brown, W. H., and Poon, T. *Introduction to Organic Chemistry*, 3rd edn, Wiley, U.S.A., p. 604, ISBN 978-0-471-44451-0.
7. Miękoś, E., Klepka, T., Zieliński, M., Sroczyński, D., Garbacz, Ł., Fenyk, A., and Łukawska, A. Effect of batched water exposed to a constant magnetic field on the properties of concrete filled with waste fly ash, phosphogypsum and starch. *Polimery/Polymers*, 2022, **67**(2), 53–60.
8. Watcharakitti, J., Win, E. E., Nimnuan, J., and Smith, S. M. Modified starch-based adhesives: A review. *Polymers*, 2022, **14**(10), 2023, https://doi.org/10.3390/polym14102023.
9. Heinrich, L. A. Future opportunities for bio-based adhesives: Advantages beyond renewability. *Green Chemistry*, 2019, **21**, 1866–1888.
10. Huicochea, E. F., and Villalobos, R. R. (Eds.). *Introductory Chapter: Starch Modifications*. InTechOpen. 2018, https://doi.org/http://dx.doi.org/10.5772/intechopen.78366.
11. Woggum, T., Sirivongpaisal, P., and Wittaya, T. Properties and characteristics of dual-modified rice starch based biodegradable films. *International Journal of Biological Macromolecules*, 2014, **67**, 490–502.

12. Sikandar, M. A., Wazir, N. R., Khan, A., Nasir, H., Ahmad, W., and Alam, M. Effect of various anti-washout admixtures on the properties of non-dispersible underwater concrete. *Construction and Building Materials*, 2020, **245**, 118469.
13. Kulshreshtha, Y., Schlangen, E., Jonkers, H. M., Vardon, P. J., and Van Paassen, L. A. CoRncrete: A corn starch-based building material. *Construction and Building Materials*, 2017, **154**, 411–423.
14. Mansour, G., Zoumaki, M., Tsongas, K., and Tzetzis, D. Starch-sandstone materials in the construction industry. *Results in Engineering*, 2020, **8**, 100182.
15. Peschard, A., Govin, A., Grosseau, P., Guilhot, B., and Guyonnet, R. Effect of polysaccharides on the hydration of cement paste at early ages. *Cement and Concrete Research*, 2004, **34**(11), 2153–2158, https://doi.org/10.1016/j.cemconres.2004.04.001.
16. Peschard, A., Govin, A., Pourchez, J., Fredon, E., Bertrand, L., Maximilien, S., and Guilhot, B. Effect of polysaccharides on the hydration of cement suspension. *Journal of the European Ceramic Society*, 2006, **26**(8), 1439–1445.
17. Patural, L., Marchal, P., Govin, A., Grosseau, P., Ruot, B., and Devès, O. Cellulose ethers influence on water retention and consistency in cement-based mortars. *Cement and Concrete Research*, 2011, **41**(1), 46–55.
18. Vieira, M. C., Klemm, D., Einfeldt, L., and Albrecht, G. Dispersing agents for cement based on modified polysaccharides. *Cement and Concrete Research*, 2005, **35**(5), 883–890, https://doi.org/10.1016/j.cemconres.2004.09.022.
19. Choi, Y. C., Cho, Y. K., Shin, K. J., and Kwon, S.-J. Development and application of microcapsule for cement hydration control. *KSCE Journal of Civil Engineering*, 2016, **20**(1), 282–292.
20. Yan, Y., Wang, R., Wang, W., Yu, C., and Liu, J. Effect of starch-based admixtures on the exothermic process of cement hydration. *Construction and Building Materials*, 2021, **289**, 122903.
21. Smith, B. J., Roberts, L. R., Funkhouser, G. P., Gupta, V., and Chmelka, B. F. Reactions and surface interactions of saccharides in cement slurries. *Langmuir*, 2012, **28**(40), 14202–14217, https://doi.org/10.1021/la3015157.
22. Yan, Y., Scrivener, K. L., Yu, C., Ouzia, A., and Liu, J. Effect of a novel starch-based temperature rise inhibitor on cement hydration and microstructure development: The second peak study. *Cement and Concrete Research*, 2021, **141**, 106325.

23. Zhang, H., Feng, P., Li, L., and Wang, W. Effects of starch-type polysaccharide on cement hydration and its mechanism. *Thermochimica Acta*, 2019, **678**, 178307.
24. Zhang, H., Wang, W., Li, Q., Tian, Q., Li, L., and Liu, J. A starch-based admixture for reduction of hydration heat in cement composites. *Construction and Building Materials*, 2018, **173**, 317–322.
25. Afroz, S., Manzur, T., Borno, I. B., Hasanuzzaman, M., and Hossain, K. M. A. Potential of starch as organic admixture in cementitious composites. *Journal of Materials in Civil Engineering*, 2021, **33**(2), 04020449.
26. Afroz, S., Manzur, T., and Anwar Hossain, K. M. Arrowroot as bio-admixture for performance enhancement of concrete. *Journal of Building Engineering*, 2020, **30**, 101313.
27. Sybis, M., and Konował, E. Influence of modified starch admixtures on selected physicochemical properties of cement composites. *Materials*, 2022, **15**, 7604, https://doi.org/10.3390/ma15217604.
28. Sybis, M., Konował, E., and Prochaska, K. Dextrins as green and biodegradable modifiers of physicochemical properties of cement composites. *Energies*, 2022, **15**, 4115, https://doi.org/10.3390/en15114115.
29. Liu, J., Yu, C., Shu, X., Ran, Q., and Yang, Y. Recent advance of chemical admixtures in concrete. *Cement and Concrete Research*, 2019, **124**, 105834.
30. Joshi, S., Goyal, S., and Reddy, M. S. Corn steep liquor as a nutritional source for biocementation and its impact on concrete structural properties. *Journal of Industrial Microbiology & Biotechnology*, 2018, **45**, 657–667.
31. Le, A. T., Gacoin, A., Li, A., Mai, T. H., Rebay, M., and Delmas, Y. Experimental investigation on the mechanical performance of starch-hemp composite materials. *Construction and Building Materials*, 2014, **61**, 106–113.
32. Sandrine, U. B., Isabelle, V., Hoang, M. T., and Chadi, M. Influence of chemical modification on hemp-starch concrete. *Construction and Building Materials*, 2015, **81**, 208–215.
33. Joseph, S. K., and Xavier, A. S. Effect of starch admixtures on fresh and hardened properties of concrete. *International Journal of Scientific Engineering and Research (IJSER)*, 2016, **4**(3), 27–30.
34. Abd, S. M., Hamood, Q. Y., Khamees, A. S., and Ali, Z. H. Effect of using corn starch as concrete admixture. *International Journal of Engineering Research and Science Technology*, 2016, **5**(3), 35–44.

35. Akindahunsi, A. A. Investigation into the use of extracted starch from cassava and maize as admixture on the creep of concrete. *Construction and Building Materials*, 2019, **214**, 659–667.

36. Oni, D. O., Mwero, J., and Kabubo, C. Experimental investigation of the physical and mechanical properties of cassava starch modified concrete. *The Open Construction and Building Technology Journal*, 2019, **13**, 331–343.

37. Kone, B., Mwero, J. N., and Ronoh, E. K. Experimental effect of cassava starch and rice husk ash on physical and mechanical properties of concrete. *International Journal of Engineering Trends and Technology*, 2022, **70**(2), 343–350.

38. Nugroho, F. K. B., Sumarni, S., Thamrin, A. G., Roemintoyo, and Isnantyo, F. D. Using baking powder as additional concrete material. *IOP Conference Series: Materials Science and Engineering*, 2020, **858**(1), 012005.

39. Ibrahim, E., Isik, M., and Hulusi, O. Utilization of polysaccharides as viscosity modifying agent in self-compacting concrete. *Construction and Building Materials*, 2014, **72**, 239–247.

40. Schmidt, W., Brouwers, H. J. H., Kühne, H.-C., and Meng, B. Interactions of polysaccharide stabilising agents with early cement hydration without and in the presence of superplasticizers. *Construction and Building Materials*, 2017, **139**, 584–593.

41. Tutal, A., Partschefeld, S., Schneider, J., and Osburg, A. Effects of bio-based plasticizers, made from starch, on the properties of fresh and hardened metakaolin-geopolymer mortar: Basic investigations. *Clays and Clay Minerals*, 2020, **68**(5), 413–427.

42. Akindahunsi, A. A., and Uzoegbo, H. C. Strength and durability properties of concrete with starch admixture. *International Journal of Concrete Structures and Materials*, 2015, **9**(3), 323–335.

43. Oni, D., Mwero, J., and Kabubo, C. The effect of cassava starch on the durability characteristics of concrete. *The Open Civil Engineering Journal*, 2020, **14**, 289–301.

44. Rahman, M. T., Mohajerani, A., and Giustozzi, F. Recycling of waste materials for asphalt concrete and bitumen: A review. *Materials*, 2020, **13**, 1495, DOI:10.3390/ma13071495.

45. Glenn, G. M., Gray, G. M., Orts, W. J., and Wood, D. W. Starch-based lightweight concrete: Effect of starch source, processing method, and aggregate geometry. *Industrial Crops and Products*, 1999, **9**, 133–144.

46. Kolaian, J. H., and Park, J. H. Chemically oxidized starch dispersant. Patent US 3493561, 1970.

47. Naka, A., and Kanae, F. Viscosity reducer for highly concentrated coke/water slurry. Patent JP 59091196, 1984.

48. Yamaguchi, T., and Nishimura, Y. Dispersant having good dispersibility and judgement thereof. Patent JP 2241534, 1990.

49. Reddy, B. R., and Eoff, L. S. Biodegradable dispersants for cement compositions and methods of cementing in subterranean formations. Patent US 2004206502, 2004.

50. Cheng, F., and Hou, G. L. Synthesis and properties of sulfonated starch as superplasticizer. *Fine Chemicals*, 2006, **23**(7), 711–716.

51. He, H., Chen, Z., Wu, J., Zhou, X., and Zhu, L. Advances in concrete water-reducing agents based on natural polymers. *Gaofenzi Cailiao Kexue Yu Gongcheng/Polymeric Materials Science and Engineering*, 2016, **32**(11), 178–183.

52. Zhang, D.-F., Ju, B.-Z., Zhang, S.-F., He, L., and Yang, J.-Z. The study on the dispersing mechanism of starch sulfonate as a water-reducing agent for cement. *Carbohydrate Polymers*, 2007, **70**(4), 363–368.

53. Moeini, M. A., Hosseinpoor, M., and Yahia, A. 3D printing of cement-based materials with adapted buildability. *Construction and Building Materials*, 2022, **337**, 127614.

54. Ivaniuk, E., Ivanova, I., Sokolov, D., Tošić, Z., Eichenauer, M. F., Lordick, D., and Mechtcherine, V. Application-driven material design of printable Strain Hardening Cementitious Composites (SHCC). *Materials*, 2022, **15**, 1631, https://doi.org/10.3390/ma15051631.

Chapter 10

Properties of Innovative and Eco-friendly Pavement and Structural Materials Using Waste Marble Sludge Dust, Rice Husk Ash, Coir Fiber, Starch-Based Binder, and Carbon Quantum Dots

Mehmet Serkan Kirgiz,[a] Jamal Khatib,[b] Muhammed Syarif,[c] Erniati Bachtiar,[d] Olatokunbo M. Ofuyatan,[e] Touhami Tahenni,[f] Krishnaraj Ramaswamy,[g, h] Ahmed Ashteyat,[i] Ayub Elahi,[j] Khairunisa Muthusamy,[k] and Gideon O. Bamigboye[l]

[a]*Academia of ACMCEN, Beaverton, OR 97075, USA*
[b]*Department of Civil Engineering, Beirut Arab University, Lebanon*
[c]*Muhammadiyah Makassar University of Indonesia, IDN*
[d]*Department of Civil Engineering, Fajar University, Makassar, South Sulawesi, Indonesia*
[e]*Department of Civil Engineering, Faculty of Engineering, Covenant University, Nigeria*
[f]*Laboratory of Acoustics and Civil Engineering, Djilali Bounaama University of Khemis Miliana, Khemis Miliana, Algeria*
[g]*Centre for Excellence-Indigenous Knowledge, Innovative Technology Transfer and Entrepreneurship, DambiDollo University, Ethiopia*
[h]*Department of Mechanical Engineering, College of Engineering and Technology, DambiDollo University, Ethiopia*
[i]*Department of Civil Engineering, University of Jordan, Amman 11942, Jordan*

Handbook of Innovative Adhesive Technology: Biodegradable and Cost-Effective Binder for Built Environment and Engineering Sciences
Edited by Mehmet Serkan Kirgiz

ISBN 978-981-5129-51-9 (Hardcover), 978-1-003-59892-3 (eBook)
www.jennystanford.com

[j]*Civil Engineering Department, University of Engineering and Technology, Taxila, Pakistan*
[k]*Faculty of Civil Engineering Technology, Universiti Malaysia Pahang, Lebuhraya Tun Razak, 26300 Gambang, Pahang, Malaysia*
[l]*Department of Civil Engineering, First Technical University, Ibadan, Oyo State, Nigeria*
nakres42@yahoo.com

This chapter aims to evaluate the properties of innovative and eco-friendly pavement and structure materials using waste marble sludge dust (WMSD), rice husk ash (RHA), coir fibers (CF), starch-based biodegradable binders (S-BDBs), and carbon quantum dots (CQDs). To this end, many innovative composite groups and control composites were prepared. Of them, the first group was the control composites made of S-BDBs. The second group included S-BDBs, WMSD, and CQDs. The third group was prepared with S-BDBs, RHA, and CQDs. The last group consisted of S-BDBs, WMSD, RHA, CF, and CQDs. For making those innovative composites, the S-BDB was heated with water at 198°C for 30 min. The fresh and hardened properties determined on the innovative and control groups of composites were setting time, rheology with slump, density, compressive strength, and tensile splitting strength as required by current international standards for cement-based binders. The results showed that the insertion of CQDs (2.5%) accelerates the setting time of the innovative composites, reduces the density and slump, as well as provides a slight increase in the compressive strength and splitting tensile strength after 28 days. According to the results for strength, the innovative composite containing 70% S-BDB, 30% WMSD, and 2.5% CQDs (S-BDB70%WMSD30%CQD2.5%) and the innovative composite made of 70% S-BDB, 15% WMSD, 15% RHA, 1% CF, and 2.5% CQDs (S-BDB70%WMSD15%RHA15%CF1%CQD2.5%) showed the greatest compressive strength and splitting tensile strength, respectively. It was concluded that CQDs and CF are the most beneficial admixture for the innovative composite bonded by the S-BDBs.

10.1 Introduction

The construction industry has played a major role in the economy globally [1]. Nevertheless, it seems that the construction industry has reached a point of significant change, with large-scale construction and infrastructure development pointing to an increase in cement production over several years. In 2019, Pheng and Hou suggested that the construction industry may have played a role in contributing to different greenhouse gas (GHG) emissions [2]. The two most challenging issues of recent times are environmental degradation and global warming [3–5]. It has been suggested that cement making may be a significant source of GHG emissions, which could contribute to various pollution problems. It is becoming increasingly clear that the melting of the ice cap is a consequence of increased GHG emissions in the Arctic and Antarctic regions [2]. Given that cement is a significant contributor to the world's total carbon dioxide emissions, it seems clear that the necessity of infrastructure cannot be ignored. It would be beneficial to explore the possibility of replacing cement with a material that has a lower carbon footprint while retaining or even surpassing the qualities of cement [6–9].

In order to address the GHG emissions and other forms of pollution, it may be beneficial to consider supplementary cementitious materials (SCMs) that have the potential to mimic the properties of ordinary Portland cement (OPC), thereby reducing carbon dioxide emissions [10]. It is important to note that SCMs are primarily industrial wastes and by-products. If these materials are not handled with care, they can potentially lead to various forms of pollution. SCMs with biodegradable binders, such as starch, have the potential to be used in a variety of combinations and proportions of by-products and cement to achieve the targeted properties of OPC. It might be helpful to consider a few examples of SCMs, such as fly ash (FA), corncob ash (CCA), sugarcane bagasse ash (SCBA), RHA, and ground granulated blast furnace slag (GGBFS) [11–14]. One of the SCMs extracted from rice crop agricultural waste is RHA and WMSD. The RHA contains more than 90% silicon oxide (SiO_2), while the WMSD consists of more than 60% calcium oxide (CaO). These materials could potentially be used as a silica-calcium-containing material to synthesize innovative and eco-friendly concrete. In 2012, Ramasamy *et al.* [15] conducted an investigation into RHA-containing concrete, which demonstrated a notable increase in compressive strength during the early curing stages. Nevertheless, it was observed that the compressive strength was reduced when the quantity of RHA in the combination exceeded 15% [15]. The compressive strength of concrete containing RHA was found to be enhanced by 9% to 18% in comparison to the control mixture [16]. In a similar vein, Chao-Lung *et al.* [17] found that the compressive strength of concrete containing RHA was between 1.2 and 1.5 times that of a control mixture. It may, therefore, be concluded that RHA is a cementitious material that can be successfully used. RHA emits a negligible amount of carbon dioxide [17] and has the potential to be used as structural concrete, as it has a strong bond [18], effective corrosion performance [19], good adhesion strength [20], and a relatively high compressive strength [21]. Thus, it contributes not only to the concrete's strength but also to its long-term durability, in terms of factors such as porosity, water absorption, permeability [22, 23], shrinkage [24], and creep [25]. Furthermore, several studies have been carried out to ascertain the strength of RHA concrete and WMSD concrete using a variety of cement types [10, 12]. Concrete materials with varying amounts of cement were prepared, cured, and applied to structures under the same curing conditions. However, it should be noted that the curing and construction processes required to determine the strength of concrete materials are costly and time-consuming [26]. To save money and time, researchers concentrated on developing alternative materials to optimize the mixture of concrete.

10.2 Applications of Pavement and Structure Materials Containing S-BDB, RHA, WMSD, CF, and CQDs

This section will examine the functional properties of starch currently available on the industrial market. It is typically derived from corn, cassava, potato, banana, and other cereals, which are frequently subjected to physical and chemical modifications to meet the demands of various industries [27]. Starch has been subjected to modification processes to enhance its applicability [28]. The chemical modification of starch typically entails the esterification, etherification, and oxidation of the accessible hydroxyl groups present on the α-D-glucopyranosyl units that constitute the starch polymers. The chemical modification will enhance the stability and film-forming properties of partially degraded starches, which will be employed in paper surface sizing, textile warp sizing, and adhesives [29]. The characteristics and properties of starch affect the efficiency of the binder. Starch binders have been observed to exhibit good ductility, good adhesion properties, self-curing properties, and hygroscopicity resistance. The term "hygroscopicity resistance" refers to the inability of the binder to absorb water, which prevents the occurrence of swelling. To enhance the structural integrity and binding capacity of starch as a primary binder, it can be subjected to modification through physical or chemical methods.

10.3 Chemical Properties of S-BDB, RHA, WMSD, CF, and CQDs

It should be noted that the chemical composition of S-BDB, RHA, WMSD, CF, and CQDs can vary depending on the source from which they are derived. The chemical compositions of the materials (S-BDB, RHA, WMSD, CF, and CQDs) were determined through X-ray fluorescence analysis [30]. The chemical compositions of various S-BDB, RHA, WMSD, CF, and CQDs materials utilized in numerous studies are presented in Table 10.1.

The predominant element present in the WMSD and RHA is calcium oxide (CaO) and silicon oxide, respectively, which exhibit a range of 29–55% and 82–94%.

10.4 Physical Properties of S-BDB, RHA, WMSD, CF, and CQDs

A summary of the physical properties of S-BDB, RHA, WMSD, CF, and CQDs used is provided in Table 10.2.

Table 10.1 Chemical compositions of various S-BDB, RHA, WMSD, CF, and CQD materials utilized in numerous studies

Types of Raw Material	Chemical Properties (%)								
	CaO	SiO_2	Al_2O_3	Fe_2O_3	MgO	SO_3	K_2O	Na_2O	Amylose
S-BDB	0.45–0.5	—	—	—	0.15–0.2	0.8–0.9	0.5–0.6	0.4–0.5	23
RHA	0.48–2.9	82–94	0.04–1.5	0.15–3.65	0.28–2.1	0.04–0.5	0.5–3.8	0.05–1.1	—
WMSD	29–55	0.18–29	0.2–10.8	0.04–10	0.2–23	0.01–0.25	0.03–4.43	0.06–3.73	—
CF	—	—	—	—	—	—	—	—	—
CQD	—	—	—	—	—	—	—	—	—
	Chemical Properties (%)								
Types of Raw Material	Amylopectin	Lignin	C	Holocellulose	α-cellulose	Pectin	Alkali	Ash	LOI
S-BDB	72	14	—	—	—	—	—	1.7–1.9	0.8
RHA	—	—	—	—	—	—	—	—	—
WMSD	—	—	—	—	—	—	—	—	—
CF	—	—	—	8–9	60–68	14–15	3–5	1–1.5	1–1.5
CQD	—	—	99.9	—	—	—	—	—	0.1

Table 10.2 A summary of the physical properties of S-BDB, RHA, WMSD, CF, and CQDs used

Types of Raw Material	Physical Properties			
	Mean Particle Size (μm)	Specific Gravity (g/cm^3)	Fineness, Passing 45 μm (%)	Specific Surface Area (m^2/g)
RHA	3–64	2–2.12	99	2–37
WMSD	1.3–9	2.53	98.7	0.62
CF	—	—	—	—
CQD	—	2.23	100	1200

10.5 Preparation and Testing of Innovative Composite and Control Composite

To evaluate the properties of innovative and eco-friendly pavement and structural materials using WMSD, RHA, CF, starch-based binder, and CQDs, a series of innovative composite and control composite groups were prepared. The first was the control composite group made of S-BDBs only. The second group included mixed S-BDBs, WMSD, and CQDs. The third group was prepared with S-BDBs, RHA, and CQDs. Besides, the last group consisted of S-BDBs, WMSD, RHA, CF, and CQDs. To conduct extensive tests on the fresh and hardened states of the innovative composite groups and control composite groups, it was decided that the S-BDBs should be cooked with water at the ratio of 40% by weight of the S-BDB at 198°C for 30 min. The fresh and hardened properties determined on the innovative and control groups of composites were setting time, rheology with slump, density, compressive strength, and tensile splitting strength as required by current international standards for cement-based binders.

10.6 Effect of RHA, WMSD, CF, and CQDs on the Fresh Properties of Pavement and Structural Materials Made of S-BDB

It is worth noting that several factors can potentially influence the properties of concrete, including chemical configuration, molecule weight, and functional mixing groups [31]. As new materials are investigated, there is a possibility that the specifications for high-strength and durable concrete may evolve continuously. The study aims to gain a deeper understanding of the potential benefits of various admixtures on the qualities of both fresh and hardened concrete. It is worth noting that oil-based and nonrenewable compounds with potential environmental concerns, such as formaldehyde, are commonly used in cementitious systems [32]. Given the concept of eco-friendly concrete, it seems that a bio-based admixture could be a promising alternative to an oil-based polymer admixture, offering a potential solution that is both effective and more affordable. This is why starch is considered. Crépy *et al.* [33] observed that starch is one of the most prevalent polymers in the world. Starch and its derivatives have the potential to beneficially modify viscosity. In Africa, it is estimated that around 5 million tons of rice husk are created annually. It has been observed that rice husks can harm the general appearance of the environment and may potentially pose a threat to the ecosystem [33]. Following a chemical analysis, it was found that rice waste could be an interesting pozzolana, potentially replacing some of the cement. It met the requirements of ASTM C618 [34] in terms of physicochemical characteristics. This is similar to the substitution of cement by RHA, S-BDB, and WMSD in the pioneer industry for the production of pavement and structural materials. This could help decrease the demand for cement, conserve resources, and reduce the pollution linked to cement production. Given the difficulty of waste disposal, researchers have chosen to explore other environmentally friendly materials that may be utilized in concrete manufacturing. One of the objectives of this chapter is to utilize S-BDB as a biodegradable binder and substitute WMSD and RHA with S-BDB partially in a composite using CF and CQDs as admixtures. This approach could help address one of the most serious environmental challenges in developing countries.

10.6.1 Setting Time

Figure 10.1 shows the setting times of composites containing S-BDB, RHA, WMSD, CF, and CQDs. It can be seen from the literature that there is a consensus on the acceleration of the initial setting (IS) and final setting (FS) times of the composite containing CQDs.

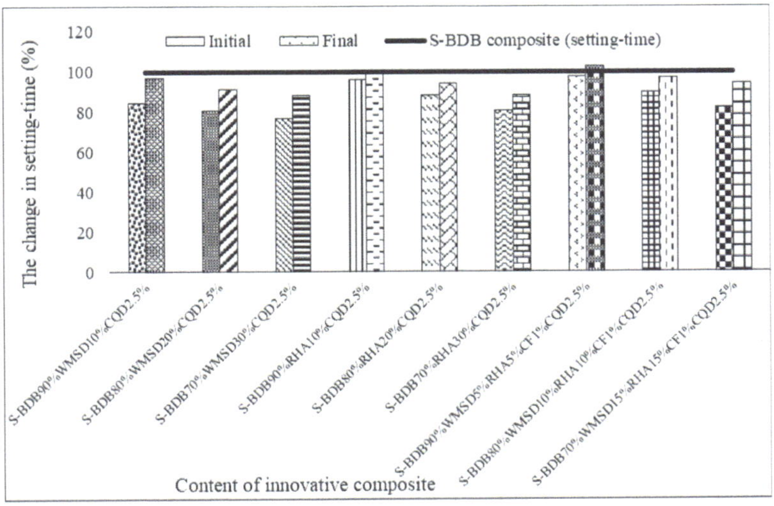

Figure 10.1 Setting times of composites containing S-BDB, RHA, WMSD, CF, and CQDs: initial and final setting times (min).

It can be seen that because of the effect of CQDs on the setting time, RHA and WMSD behaved in a way distinct from that of the literature. Concerning the impact of RHA content on the setting time of innovative composites, it is notable that there is no universal consensus in the literature. Some researchers [35, 36] have indicated that RHA may have a retarding effect [37, 38] on setting times, while others have reported the opposite [33, 39–41], with RHA having an accelerating effect. Evaluating the innovative composite with the blending of S-BDB, RHA, WMSD, CF, and CQDs, it was found that while RHA increased the delay in the setting time with an increasing quantity of RHA, WMSD decreased the delay in the setting time with an increasing quantity of WMSD in this study. Regarding ternary binders containing S-BDB, RHA, WMSD, and CQDs, this current study presents the initial and final setting times, as illustrated in Fig. 10.1. There may be a trend indicating that the binary use of RHA and WMSD with CQDs may have resulted in a notable prolongation of the initial and final setting times in comparison to the control composite containing S-BDB only. It may be the case that the greater

S-BDB content is responsible for the observed retardation in setting time. It may also be observed that the setting times for a ternary binder appear to decrease as the percentage replacement of S-BDB with RHA and WMSD increases. Brooks *et al.* suggested that the delay could be attributed to a combination of factors, including a lower binder content and the effective admixture dosage per weight of the binder [42]. Additionally, it is possible that starch polymer adsorption to the oxides of carbon from CQDs, as well as calcium, silicon, and aluminum from RHA and WMSD, which govern the set properties of starch-based binder paste, may contribute to the observed acceleration [43–45]. Of all the mixtures, the S-BDB70%-WMSD30%-CQD2.5% combination showed the highest acceleration in both initial and final setting times.

10.6.2 Rheology with Slump

There was still a decrease in the slump of innovative composites using 2.5% CQDs. The workability may have decreased somewhat, from 68 mm for the innovative composites (S-BDB-WMSD-CQD, S-BDB-RHA-CQD, and S-BDB-WMSD-RHA-CF-CQD composites) to 38 mm after the 2.5% addition of CQDs. It seems plausible to suggest that this may be linked to the thickening power of CQDs, which could result in the viscosity-modifying properties of the starch binder. The leafy shape of CQDs may allow the starch gel to easily envelop the surface of WMSD and RHA grains, which could increase the mix viscosity [32, 46]. The presence of WMSD, RHA, and CF in the mix may contribute to an increase in the number of fine particles in innovative composites, which could potentially lead to an increase in the water demand of the mix. The high demand for water in WMSD-, RHA-, and CF-based composites may be related to the porous structure of by-products, as would be understandood from the SEM analysis of WSMD, RHA, and CF in the most current literature. The slump of WSMD-, RHA-, and WSMD-RHA-CF-based innovative composites decreased in all mixes compared to the S-BDB control composite. It would appear that the lowest slump falls within the range of the mix design (10–30 mm). It seems reasonable to suggest that the decrease in the workability of WSMD-, RHA-, and WSMD-RHA-CF-based innovative composites may be due to the porous nature of WSMD, RHA, and CF, which may require more water for adequate workability. Additionally, the particle size was found to be relatively fine, which may have contributed to an increase in the quantity of fine particles present in the matrix [47]. It may also be related to the viscosity-modifying effect of CQDs in innovative composites.

10.7 Effect of RHA, WMSD, CF, and CQDs on the Hardened Properties of Pavement and Structure Materials Made of S-BDB

10.7.1 Density

There is a gradual increase in density, which reaches a maximum at 2.5% CQDs and then declines. The reduction may be due to the reduced workability of the composite, which could potentially influence and give better compaction. The density of the S-BDB control composite may have been affected by the increased percentage of WMSD, RHA, and CF at 28 days. It seems reasonable to suggest that the decrease in density may be attributed to the lower specific gravity of WMSD, RHA, and CF, which could result in a direct decrease in matrix density with an increase in WMSD, RHA, and CF concentration. The density was found to be within the range of 2200–2600 kg/m^3 for normal concrete, which aligns with the specifications outlined by the BS EN 12390-7 [48]. It is also worth noting that all the innovative composites (S-BDB-WMSD-CQD, S-BDB-RHA-CQD, and S-BDB-WMSD-RHA-CF-CQD) fall within this interval. From the literature, it can be observed that there is a progressive decrease in density across all mixes, from the composite containing 90% S-BDB, 10% WMSD, and 2.5% CQDs (S-BDB90%WMSD10%CQD2.5%) and the composite made of 70% S-BDBs, 15% WMSD, 15% RHA, 1% CF, 2.5% CQDs (S-BDB70%WMSD15%RHA15%CF1%CQD2.5%), with a range of 2474 kg/m^3 to 2359 kg/m^3. This may be due to WMSD, RHA, and CF having a lower specific gravity, which could have resulted in a direct decrease in the density of the matrix with an increase in the percentage of WMSD, RHA, and CF. This may have led to some issues with compaction, potentially creating voids in the composite. The density was found to be from 2000 to 2600 kg/m^3.

10.7.2 Compressive Strength

From the analysis of the current experimental literature, it appears that there may be a reduction in the 28th-day compressive strength when cement is replaced with WMSD and RHA by up to 20%. It was observed that the compressive strengths of innovative composites using S-BDB, WMSD, RHA, CF, and CQDs were somewhat greater than that of the S-BDB composite at 28 days. Figure 10.2 shows the change in the 28th-day compressive strength of innovative composites containing S-BDB, RHA, WMSD, CF, and CQDs and the S-BDB composite.

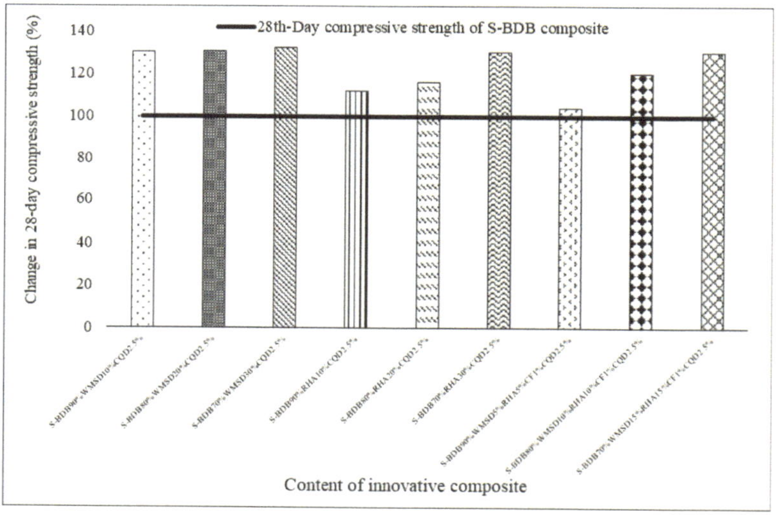

Figure 10.2 Change in the 28th-day compressive strength of innovative composites containing S-BDB, RHA, WMSD, CF, and CQDs and the S-BDB composite.

This observation is in close alignment with the findings reported by Marthong [49] as well as Kartini *et al.* [50]. It seems plausible to suggest that the significantly greater calcium oxide content of RHA (between 0.48% and 2.9%) and WMSD (between 29% and 55%) compared to starch binder (between 0.45% and 0.5%) might account for the increase in the compressive strength of innovative composites, including S-BDB-WMSD-CQD, S-BDB-RHA-CQD, and S-BDB-WMSD-RHA-CF-CQD. However, it is worth noting that at 28 days, there seems to be a slight increase in strength for all replacement levels, when compared to the S-BDB composite. It would appear that a substitution percentage of 10% is associated with the highest compressive strength, which is comparable to that of the control. This strength gain may be the result of a later hydraulic reaction of WMSD and RHA in the innovative composites [51]. The compressive strengths of composites made of S-BDB-WMSD-CQD, S-BDB-RHA-CQD, and S-BDB-WMSD-RHA-CF-CQD are greater than that of the S-BDB composite after 28 curing days. It seems reasonable to suggest that this may be linked to the increasing behavior of CQD, which has the effect of reducing bleeding and segregation, as well as an increase in bonding and a strengthening effect on the calcium oxide content in the innovative composites due to its leafy structure. Additionally, internal curing may enhance the strength gain due to the viscosity-modifying properties of starch. According to Swinkels, S-BDB has a higher degree of polymerization, which gives the material an increased binding force [52]. This could be a potential reason why using S-BDB as a binder enhances compressive strength. These results are in line with the findings of Wen *et al.* [32], Cao *et al.* [53], Adam *et al.* [54], Oni *et al.* [55], and Akindahunsi and Uzoegbo [56]. From a review of the current literature on the compressive strength of composites blended with WMSD, RHA, CF, and S-BDB, it appears that there may be a slight increase in composite strength after 28 days of curing for all percentage replacements of WMSD, RHA, CF, and S-BDB compared to the S-BDB composite, with a growth of over 130% in some cases. It is worth noting that the S-BDB-WMSD-CQD, S-BDB-RHA-CQD, and S-BDB-WMSD-RHA-CF-CQD mixes exhibited enhanced compressive strengths compared to the S-BDB composite after 28 days of curing, varying between 28 and 33.1 MPa. The increase in force may be attributed to WMSD, RHA, and CF being a latent hydraulic material that takes time to react with S-BDB and CQDs. In addition, the viscosity power of S-BDB may help avoid segregation and bleeding in composites and keep the matrix well-packed together. Thus, it could be suggested that the optimum content is at the innovative composite mix consisting of 70% S-BDB, 30% WMSD, and 2.5% CQDs (S-BDB70%WMSD30%CQD2.5%). Additionally, it should be noted that the S-BDB70%WMSD30%CQD2.5% mix has the greatest compressive strength after 28 days.

10.7.3 Splitting Tensile Strength

Figure 10.3 shows the change in the 28th-day splitting tensile strengths of innovative composites containing S-BDB, RHA, WMSD, CF, and CQDs and the S-BDB composite. The current literature on the splitting tensile strengths of composites with WMSD, RHA, and CF illustrates a pattern similar to that observed in the compressive strength results of innovative composites made of WMSD, RHA, CF, CQDs, and S-BDB. It seems that at 28 days, there is a slight decrease in the splitting tensile strengths of WMSD- and RHA-based cement concrete, accompanied by a percentage increase. It may be that the bonding behavior of WMSD and RHA in the constituent materials of cement-based concrete is not as effective as that of the S-BDB composite. Therefore, it seems that the splitting tensile strengths of innovative composites having S-BDB, WMSD, RHA, CF, and CQDs may have increased when compared to the S-BDB composite at 28 days, with the maximum splitting tensile strength observed for the innovative composite prepared with 70% S-BDB, 15% WMSD, 15% RHA, 1% CF, and 2.5% CQDs (S-BDB70%WMSD15%RHA15%CF1%CQD2.5%). It may be the case that the bonding behavior of WMSD, RHA, and

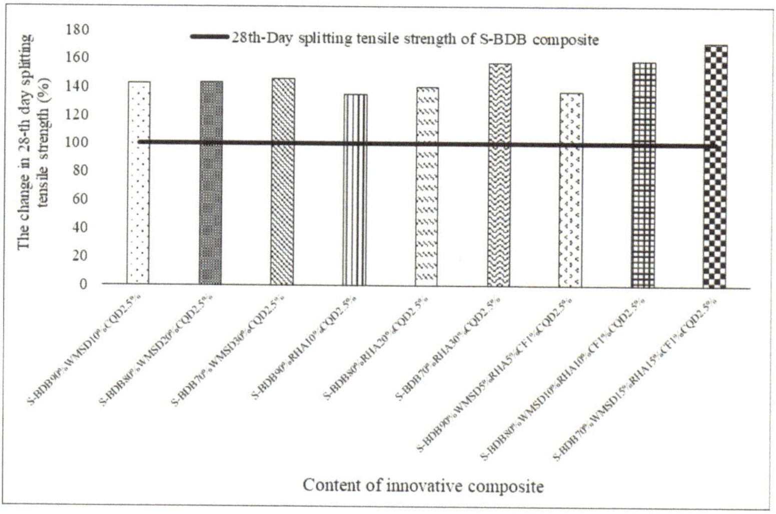

Figure 10.3 Change in the 28th-day splitting tensile strengths of innovative composites containing S-BDB, RHA, WMSD, CF, and CQDs and the S-BDB composite.

CF in the constituent materials of composites is as effective as that of cement-based traditional composites. It seems that the splitting tensile strengths of innovative composites—S-BDB-WMSD-CQD, S-BDB-RHA-CQD, and S-BDB-WMSD-RHA-CF-CQD—may have increased compared to the S-BDB composite at 28 days, with the greatest splitting tensile strength observed for both the composite with 70% S-BDB, 15% WMSD, 15% RHA, 1% CF, and 2.5% CQDs (S-BDB70%WMSD15%RHA15%CF1%CQD2.5%) and the composite with 70% S-BDB, 30% RHA, and 2.5% CQDs (S-BDB70%RHA30%CQD2.5%). It may be the case that the latent hydraulic effect of WMSD and RHA is responsible for this. It can be observed from the literature on splitting tensile strength that the addition of CF to concrete results in improved performance at all ages, with 1% CF showing the best results compared to the control.

However, increasing the amounts of WMSD and RHA may harm the splitting tensile strength, though CQDs were used as a strength gainer. A possible explanation for this could be the ability of starch to reduce bleeding in the innovative composites, as the CQDs might help minimize the formation of micro-cracks at the paste–aggregate interphase, as suggested by Shetty and Jain [57]. Regarding the effect of CQDs on the splitting tensile strengths of S-BDB innovative composites prepared with WMSD, RHA, and CF and cured for 28 days, it would appear that there is a similar trend in the splitting tensile strength as in the compressive strength. From the results in the existing literature, it seems that the percentage increase in WMSD and RHA has resulted in a reduction in the splitting tensile strength. For instance, at the age of 28 days, the reduction was observed to be 2.3% and 1.1%, respectively, for another study which inserted 10% RHA and 15% RHA, when compared to the conventional cement-based concrete. It may be observed that this improvement on the 28th day is indicative of strength development, which could be attributed to the high degree of polymerization of starch [18]. After 28 days of curing, it appears that the mix containing 70% S-BDB, 15% WMSD, 15% RHA, 1% CF, and 2.5% CQDs (S-BDB70%WMSD15%RHA15%CF1%CQD2.5%) demonstrated the highest splitting tensile strength. It is worth noting that the increased percentage of WMSD and RHA incorporation with CF and CQDs appears to have resulted in a slight increase in the splitting tensile strength. This could be attributed to the latent hydraulic activity of WMSD and RHA, which is a long-term process, as well as the strength-increasing effect of CQDs and CF, which requires time to achieve the desired strength. Furthermore, the impact of S-BDB on the ternary mixture may be optimized through its viscosity-modifying effect, which can prevent bleeding and segregation while potentially enhancing the mixture's binding forces. This is due to its ability to facilitate polymerization in micropores within the matrix [52].

10.8 Conclusion

This chapter aimed to evaluate the properties of innovative and eco-friendly pavement and structural materials using WMSD, RHA, CF, starch-based binder, and CQDs. The following inferences can be made from the chapter. CQDs accelerate the setting time of innovative composites, reduce the density and slump, and provide a slight increase in compressive strength and splitting tensile strength at 28 days. The results demonstrated that the innovative composite containing 70% S-BDB, 30% WMSD, and 2.5% CQDs (S-BDB70%WMSD30% CQD2.5%) and the innovative composite made of 70% S-BDB, 15% WMSD, 15% RHA, 1% CF, and 2.5% CQDs (S-BDB70%WMSD15%RHA15%CF1%CQD2.5%) exhibited the most notable compressive strength and splitting tensile strength, respectively. This indicates that CQDs and CF are the most advantageous admixture for the innovative composites bonded by the S-BDB, as demonstrated by the results of the chapter.

References

1. Adnan, Z.S., Ariffin, N.F., Mohsin, S.M.S., and Lim, N.H.A.S. Review paper: Performance of rice husk ash as a material for partial cement replacement in concrete. *Mater. Today Proc.*, 2021, **48**, 842–848.
2. Pheng, L.S., and Hou, L.S. The economy and the construction industry. In: *Construction Quality and the Economy. Management in the Built Environment.* Springer, Singapore, 2019, pp. 21–54.
3. Eijgelaar, E., Thaper, C., and Peeters, P. Antarctic cruise tourism: The paradoxes of ambassadorship, "last chance tourism" and greenhouse gas emissions. *J. Sustain. Tour.*, 2010, **18**(3), 337–354.

4. Arrigoni, A., Panesar, D.K., Duhamel, M., Opher, T., Saxe, S., Posen, I.D., and MacLean, H.L. Life cycle greenhouse gas emissions of concrete containing supplementary cementitious materials: Cut-off vs. substitution. *J. Clean. Prod.*, 2020, **263**.
5. Chen, C., Habert, G., Bouzidi, Y., and Jullien, A. Environmental impact of cement production: Detail of the different processes and cement plant variability evaluation. *J. Clean. Prod.*, 2010, **18**(5), 478–485.
6. Lamb, W.F., Wiedmann, T., Pongratz, J., Andrew, R., Crippa, M., Olivier, J.G.J., Wiedenhofer, D., Mattioli, G., Al Khourdajie, A., House, J., Pachauri, S., Figueroa, M., Saheb, Y., Slade, R., Hubacek, K., Sun, L., Ribeiro, S.K., Khennas, S., de la Rue du Can, S., Chapungu, L., Davis, S.J., Bashmakov, I., Dai, H., Dhakal, S., Tan, X., Geng, Y., Gu, B., and Minx, J. A review of trends and drivers of greenhouse gas emissions by sector from 1990 to 2018. *Environ. Res. Lett.*, 2021, **16**(7), 049502.
7. Farooq, F., Wisal Ahmed, W., Akbar, A., Aslam, F., and Alyousef, R. Predictive modeling for sustainable high-performance concrete from industrial wastes: A comparison and optimization of models using ensemble learners. *J. Clean. Prod.*, 2021, **292**, 126032.
8. Mahlia, T.M.I. Emissions from electricity generation in Malaysia. *Renewable Energy*, 2002, **27**(2), 293–300.
9. Akbar, A., Farooq, F., Shafique, M., Aslam, F., Alyousef, R., and Alabduljabbar, H. Sugarcane bagasse ash-based engineered geopolymer mortar incorporating propylene fibers. *J. Build. Eng.*, 2021, **33**, 101492.
10. Crossin, E. The greenhouse gas implications of using ground granulated blast furnace slag as a cement substitute. *J. Clean. Prod.*, 2015, **95**, 101–108.
11. Rathnarajan, S., Dhanya, B.S., Pillai, R.G., Gettu, R., and Santhanam, M. Carbonation model for concretes with fly ash, slag, and limestone calcined clay using accelerated and five-year natural exposure data. *Cem. Concr. Compos.*, 2022, **126**, 104329.
12. Ameri, F., Shoaei, P., Bahrami, N., Vaezi, M., and Ozbakkaloglu, T. Optimum rice husk ash content and bacterial concentration in self-compacting concrete. *Constr. Build. Mater.*, 2019, **222**, 796–813.
13. Antiohos, S.K., Papadakis, V.G., and Tsimas, S. Rice husk ash (RHA) effectiveness in cement and concrete as a function of reactive silica and fineness. *Cem. Concr. Res.*, 2014, **61–62**, 20–27.
14. Zhang, Z. Yang, F., Liu, J-C., and Wang, S. Eco-friendly high strength, high ductility engineered cementitious composites (ECC) with substitution of fly ash by rice husk ash. *Cem. Concr. Res.*, 2020, **137**, 106200.

15. Ramasamy, V. Compressive strength and durability properties of rice husk ash concrete. *KSCE J. Civil Eng.*, 2012, **16**(1), 93–102.
16. Khassaf, S.I., Jasim, A.T., and Mahdi, F.K. Investigation the properties of concrete containing rice husk ash to reduction the seepage in canals. *Int. J. Sci. Technol. Res.*, 2014, **3**(4), 348–354.
17. Chao-Lung, H., Anh-Tuan, B.L., and Chun-Tsun, C. Effect of rice husk ash on the strength and durability characteristics of concrete. *Constr. Build. Mater.*, 2011, **25**(9), 3768–3772.
18. Kusbiantoro, A., Nuruddin, M.F., Shafiq, N., and Qazi, S.A. The effect of microwave incinerated rice husk ash on the compressive and bond strength of fly ash based geopolymer concrete. *Constr. Building Materials*, 2012, **36**, 695–703.
19. Saraswathy, V., and Song, H.W. Corrosion performance of rice husk ash blended concrete. *Constr. Build. Mater.*, 2007, **21**(8), 1779–1784.
20. Basri, M.S.M., Mustapha, F., Mazlan, N., and Ishak, M.R. Optimization of adhesion strength and microstructure properties by using response surface methodology in enhancing the rice husk ash-based geopolymer composite coating. *Polymers*, 2020, **12**(11), 1–15, 2709.
21. Kishore, R., Bhikshma, V., and Jeevana Prakash, P. Study on strength characteristics of high strength Rice Husk Ash concrete. *Procedia Eng.*, 2011, **14**, 2666–2672.
22. Ganesan, K., Rajagopal, K., and Thangavel, K. Rice husk ash blended cement: Assessment of optimal level of replacement for strength and permeability properties of concrete. *Constr. Build. Mater.*, 2008, **22**(8), 1675–1683.
23. De Souza Rodrigues, C., Ghavami, K., and Stroeven, P. Porosity and water permeability of rice husk ash-blended cement composites reinforced with bamboo pulp. *J. Mater. Sci.*, 2006, **41**(21), 6925–6937.
24. Habeeb, G.A., and Fayyadh, M.M. Rice husk ash concrete: The effect of RHA average particle size on mechanical properties and drying shrinkage. *Aust. J. Basic Appl. Sci.*, 2009, **3**(3), 1616–1622.
25. Muhammad, A., Thienel, K. C., and Sposito, R. Suitability of blending rice husk ash and calcined clay for the production of self-compacting concrete: A review. *Materials*, 2021, **14**(21).
26. Rattanachu, P., Toolkasikorn, P., Tangchirapat, W., Chindaprasirt, P., and Jaturapitakkul, C. Performance of recycled aggregate concrete with rice husk ash as cement binder. *Cem. Concr. Comp.*, 2020, **108**, 103533.
27. Waliszewski, K.N., Aparicio, M.A., Bello, L.A., and Monray, J.A. Changes of banana starch by chemical and physical modification. *Carbohydr. Polym.*, 2003, **52**, 237–242.

28. Adak, S., and Banerjee, R. A green approach for starch modification: Esterification by lipase and novel imidazolium surfactant. *Carbohydr. Polym.*, 2016, **150**, 359–368.

29. Chiu, C.W., and Solarek, D. Modification of starches. *Food Sci. Technol.*, 2009, **3**, 629–655.

30. American Society of Testing and Materials. ASTM C 114-18e. Standard test methods for chemical analysis of hydraulic cement. *ASTM Int.*, 2018, 1–33.

31. Liu, J., Zhou, H., and Ouyang, P. Effect of straw mixing amount on mechanical properties of admixture-adding hollow block. *J. Wuhan Univ. Technol. Mater. Sci. Ed.*, 2013, **28**(3), 508–513,

32. Wen, Q., Li, C., Cai, Z., Zhang, W., Gao, H., Chen, L., Zeng, G., Shu, X., and Zhao, Y. Study on activated carbon derived from sewage sludge for adsorption of gaseous formaldehyde. *Bioresour. Technol.*, 2011, **102**(2), 942–947.

33. Crépy, L., Miri, V., Joly, N., Martin, P., and Lefebvre, J.M. Effect of side-chain length on the structure and thermomechanical properties of fully substituted cellulose fatty esters. *Carbohydr. Polym.*, 2011, **83**(4), 1812–1820.

34. ASTM C618. Standard specification for coal fly ash and raw or calcined natural pozzolan for use in concrete. *ASTM Int.*, 2005.

35. Ganesan, K., Rajagopal, K., and Thangavel, K. Rice husk ash blended cement: Assessment of optimal level of replacement for strength and permeability properties of concrete. *Constr. Build. Mater.*, 2008, **22**(8), 1675–1683.

36. Krishna, N.K., Sandeep, S., and Mini, K.M. Study on concrete with partial replacement of cement by rice husk ash. *IOP Conf. Ser. Mater. Sci. Eng.*, 2016, **149**(1), 012109.

37. Givi, A.N., Rashid, S.A., Aziz, F.N.A., and Salleh, M.A.M. Contribution of rice husk ash to the properties of mortar and concrete: A review. *J. Am. Sci.*, 2010, **6**(3), 157–165.

38. El-Dakroury, A., and Gasser, M.S. Rice husk ash (RHA) as cement admixture for immobilization of liquid radioactive waste at different temperatures. *J. Nucl. Mater.*, 2008, **381**(3), 271–277.

39. BS EN 933-1. The particle size distribution of aggregate. *British Standards Institution*, 2012.

40. BS EN 196-2. Chemical properties of cement. *British Standards Institution*, 2013.

41. BS 8500-2. The mix design. *British Standards Institution*, 2012.

42. Brooks, J.J., Johari, M.A.M., and Mazloom, M. Effect of admixtures on the setting times of high-strength concrete. *Cem. Concr. Compos.*, 2000, **22**(4), 293–301.
43. BS EN 196-3. Setting time and soundness of cement. *British Standards Institution*, 2016.
44. Popova, A., Geoffroy, G., Renou-Gonnord, M.F., Faucon, P., and Gartner, E. Interactions between polymeric dispersants and calcium silicate hydrates. *J. Am. Ceram. Soc.*, 2000, **83**(10), 2556–2560.
45. Tsao, K.H. FIMP dark matter freeze-in gauge mediation and hidden sector. *J. Phys. G Nucl. Part. Phys.*, 2018, **45**(7), 830–842.
46. Akindahunsi, A.A., and Schmidt, W. Effect of cassava starch on shrinkage characteristics of concrete. *African J. Sci. Technol. Innov. Dev.*, 2019, **11**(4), 441–447.
47. Zaid, O., Ahmad, J., Siddique, M.S., and Aslam, F. Effect of incorporation of rice husk ash instead of cement on the performance of steel fibers reinforced concrete. *Front. Mater.*, 2021, **8**, 1–14, 665625.
48. BS EN 12390-7. Determining the density of hardened concrete. *British Standards Institution*, 2019.
49. Marthong, C. Effect of rice husk ash (RHA) as partial replacement of cement on concrete properties. *Int. J. Eng. Res. Technol.*, 2016, **1**(6), 1–3.
50. Kartini, K., Nazierah, M.N., Zaidahtulakmal, M., and Aisyah, G.S. Effects of silica in rice husk ash (RHA) in producing high strength concrete. *Int. J. Eng. Technol.*, 2012, **2**(12), 1951–1956.
51. Chindaprasirt, P., and Rukzon, S. Strength, porosity and corrosion resistance of ternary blend Portland cement, rice husk ash and fly ash mortar. *Constr. Build. Mater.*, 2008, **22**(8), 1601–1606.
52. Swinkels, J.J.M. Composition and properties of commercial native starches, *Starch Stärke*, 1985, **37**(1), 1–5.
53. Cao, F., Qiao, H., Li, Y., Shu, X., and Cui, L. Effect of highland barley straw ash admixture on properties and microstructure of concrete. *Constr. Build. Mater.*, 2022, **315**, 125802.
54. Adam, M.G.O., Koteng, D.O., Thuo, J.N., and Matallah, M. Analysing the effect of cassava flour as a mixture on the physical, mechanical, and durability properties of high-strength concrete. *Civil Eng. J.*, 2022, **8**(12), 3866–3882.
55. Oni, O.D., Mwero, J., and Kabubo, C. Experimental investigation of the physical and mechanical properties of cassava starch modified concrete. *Open Constr. Build. Technol. J.*, 2020, **13**(1), 331–343.

56. Akindahunsi, A.A., and Uzoegbo, H.C. Strength and durability properties of concrete with starch admixture. *Int. J. Concr. Struct. Mater.*, 2015, **9**(3), 323–335.

57. Shetty, M.S., and Jain A.K. *Concrete Technology: Theory and Practice.* Chand (S.) & Co Ltd. 2019.

Chapter 11

Behavior of Bacterial Strain over Cementless Geopolymer Concrete for Sustainable Development

A. Chithambar Ganesh,[a] M. Vinod Kumar,[a] A. Navya,[b] G. Dasaradha Rami Reddy,[b] S. R. Ragavi,[b] C. Navin Gowtham,[b] and Mehmet Serkan Kirgiz[c]

[a]*Vel Tech Rangarajan Dr. Sagunthala R&D Institute of Science and Technology, India*

[b]*Sree Vidyanikethan Engineering College, India*

[c]*Academia of ACMCEN, Beaverton, OR 97075, USA*

nakres42@yahoo.com

Many research works have thrown light on the utilization of bacteria in cement concrete, but only a few have investigated the influence of bacterial strain in the geopolymer concrete (GPC). GPC utilized in this discussion is based on ground granulated blast furnace slag (GGBS) and hence was cured under ambient conditions only. The key research objective of the current work investigates the effect of *Bacillus subtilis* and *Bacillus sphaericus* over the GPC in mechanical properties. The real contribution of bacterial strain in concrete

Handbook of Innovative Adhesive Technology: Biodegradable and Cost-Effective Binder for Built Environment and Engineering Sciences
Edited by Mehmet Serkan Kirgiz

ISBN 978-981-5129-51-9 (Hardcover), 978-1-003-59892-3 (eBook)
www.jennystanford.com

can be witnessed in the self-healing property. Artificial cracks are induced in the GPC, and the self-healing mechanism is illustrated through the intermittent photographs. The findings claim the significant presence of bacterial precipitation, which heals the cracks completely. The research findings lay the path for the sustainable development of building materials.

11.1 Introduction

The word "geopolymer" was formed by Professor Joseph Davidovits in 1978 to show a wide variety of substances distinguished by systems of artificial molecules. Geopolymers depend on thermally activated organic substances such as metakaolin or industrial by-products such as slag or fly ash to offer an active source of silicon and aluminum [1]. Si and Al are disintegrated in an activator solution and polymerized into molecular chains, forming the matrix. GPC is a form of concrete made by operating aluminate- and silicate-bearing substances with an acoustic activator. Generally, unused substances such as slag or fly ash from iron and metal goods are utilized, which leads to hygienic surroundings [1]. The constituents of GPC are binder materials, Na_2SiO_3, NaOH solid, water (to dilute NaOH), and fine and coarse aggregates. The main parameters that control the properties of GPC are a type of binder material, alkaline binder ratio, and molarity of alkaline solution [2, 3]. Geopolymers can be manufactured from fly ash [4–6], GGBS [7, 8], wood ash [9], waste glass powder [10], biomedical waste ash [11], rice husk ash [12, 13], and metakaolin [14, 15]. GGBS-based GPC needs no heat curing, and it could be applied as bricks, paver blocks, and precast slabs [9]. The GPC exhibits higher compressive strength, flexural strength, and resistance to acid attack. The only limitation of GPC is its brittle nature and development of minor cracks at higher loads. GPC exhibits fair properties even at elevated temperatures [6]. Generally, concrete is prone to cracks, which gives less durability, strength, and serviceability. Hence, self-healing methods are used to overcome those aspects. Among them, the biological self-healing process is more efficient and gives promising results than the others. Until now, so many bacteria such as *B. subtilis* and *B. sphaericus* have been used in self-healing concrete. The Bacillus type of bacteria enhances the engineering properties of concrete; however, they can sometimes be susceptible to harsh environmental conditions, such as extreme pH levels and high temperatures.

Self-healing methods are used in major structures such as the foundation of pillars, bridges, and all types of roads. This is because these structures experience heavy loads every day, which leads to minor cracks. The use of self-healing concrete will reduce the cost of maintenance as it does not require frequent maintenance.

Concrete can be healed using different techniques such as the utilization of microcapsules with alkali activators or repairing agents, shape memory polymers, and bacteria. Self-healing can be achieved using bacteria inside concrete, which increases the service life of a material and leads to designing a sustainable infrastructure. Different bacterial strains such as *Bacillus pasteurizing, Bacillus cohnii, Bacillus balodurans, Bacillus pseudofirmus, Escherichia coli, Bacillus megaterium, Bacillus pasteurii, Bacillus cereus, B. subtilis,* and *B. sphaericus* can be used inside concrete [4, 8, 16–18].

This chapter provides a comparison of the influence of *B. subtilis* and *B. sphaericus* bacteria inside GPC. Mechanical tests on properties such as compressive, tensile, and flexural strengths are performed. Further, the self-healing ability of concrete is also investigated by imposing a 1 mm crack into the concrete and observing it for 28 days. Intermittent photographs revealed the active performance of concrete in self-healing.

11.2 Materials

GPC was prepared using GGBS as the binding material. A mixture of NaOH and Na_2SiO_3 was utilized as the activator solution. For fine aggregates (FA), M sand was used, and GGBS was obtained as an industrial waste of iron and steel ore in blast furnace. The specific gravity of GGBS was 3.01. M sand conforming to IS-383-2016 Zone 3 was used. The specific gravity of fine aggregates was 2.74. Coarse aggregates (CA) having a maximum size of 20 mm was used for the work, and the specific gravity was 2.88. Locally available water was used for the work, which conformed to IS-456-2000. The alkaline solution used was a mixture of NaOH and Na_2SiO_3 in the ratio of 1:2.5. A sodium hydroxide (NaOH) solution of 13 M molarity was used.

Na_2SiO_3 was used in the form of a gel. *B. subtilis* and *B. sphaericus* were the two strains of bacteria used in a concentration of 10^5 cells per milliliter.

11.3 Culturing of Bacteria

Natural cultures of *B. subtilis* and *B. sphaericus* were prepared individually in Luria Broth (LB), which was formed by disintegrating fine grains of agar in distilled water at 28°C. Bacteria sterilization was done in an autoclave at a temperature of 121°C. This medium was placed in a shaking incubator for 12 h at 37°C, after which a turbid whitish-yellow bacterial solution was found. The concentration of both the bacteria was 10^5 cells per milliliter.

11.4 Mix Design and Casting

The mix design for M30 grade concrete was derived as per the B. V. Rangan [19] recommendations for the mix design of GPC, which is similar to IS-10262-2009. Table 11.1 provides the typical mixed ingredients. The chapter discusses the procedure for casting geopolymer specimens using the *B. subtilis* and *B. sphaericus* strains and investigates the outcome of the bacterial strains over properties such as compressive, splitting tensile, and flexural strengths as well as self-healing. The casting of specimens was done as follows: First,

Table 11.1 Mix proportions

S. No.	Materials	Mass (kg/m^3)
1.	GGBS	550
2.	Total alkaline solution	335.5
3.	Na_2SiO_3	239.6
4.	NaOH	95.9
5.	Water content	16.5
6.	Fine aggregates	588
7.	Coarse aggregates	1112.5

GGBS and fine aggregates (M sand) were mixed for about 3 min in a blender succeeded by the addition of course aggregates. The alkaline solution was later added along with the bacterial strain in the LB medium and mixed for about 5 min. Water was then added for workability. The casted specimens were then tested as per IS guidelines.

11.5 Results and Discussion

11.5.1 Compressive Strength

Bacterial GPC cubes can be tested in a compression testing machine (CTM) of range 2000 kN. The general changes in compressive strength with the utilization of bacteria are discussed below and is depicted in Fig. 11.1.

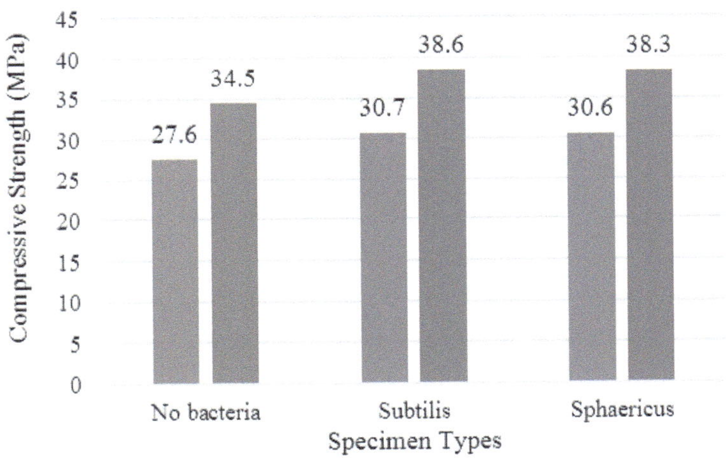

Figure 11.1 Compressive strength results for various bacterial strains.

All the GPC samples gained up to 80% of the 28-day strength in 7 days itself. With the application of bacterial strains in GPC, a rise in compressive strength of about 11% could be witnessed. It has been commonly reported that GPC specimens with the *B. subtilis* strain yielded a higher compressive strength than the *B. sphaericus*-incorporated specimens.

Ashwij *et al.* [20] claimed a 17% rise in compressive strength with the integration of the *B. subtilis* bacterial strain in cement concrete. Mutitu *et al.* [21] claimed an increase of about 17% in compressive strength with *the sphaericus* strain in the cement concrete specimens. The increase in compressive strength is caused by the precipitation of calcium silicate hydrate (CSH) in the pores due to the bacterial activity [22].

11.5.2 Splitting Tensile Strength

GPC with different bacterial strains cast in cylindrical samples were tested in a universal testing machine (UTM) of range 2000 kN. There was an increment in the tensile strength due to the bacterial contribution (Fig. 11.2).

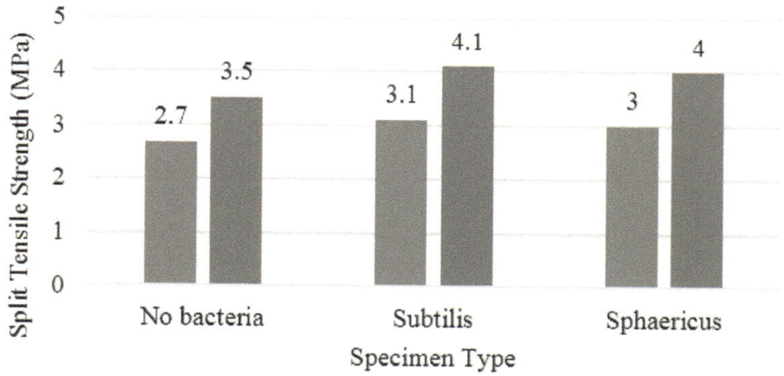

Figure 11.2 Tensile strength results for various bacterial strains.

It was explicit that geopolymer specimens with *subtilis* specimens yielded a higher increase in strength than the *sphaericus*-incorporated specimens. This depicts an increase in compressive strength. An increase of about 15% and 11% was witnessed with the *subtilis* and *sphaericus* strains in the GPC specimens in about 7 days. Priya *et al.* [22] reported a 14% rise in splitting tensile strength (STS) with the utilization of the *sphaericus* strain in engineered concrete. Meera and Subha [23] claimed a similar behavior in cement concrete. The precipitation of calcite is the explanation for the enhanced STS of the specimens.

11.5.3 Flexural Strength

Bacterial GPC prisms were tested in the 2000 kN UTM. The contribution of flexural strength through the bacterial strain is discussed below, and the changes are shown in Fig. 11.3.

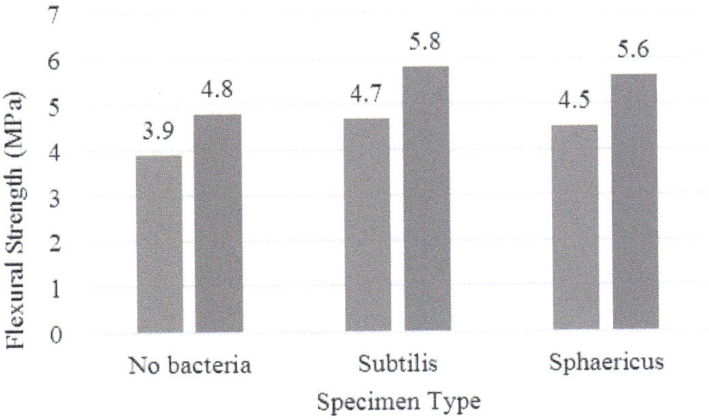

Figure 11.3 Flexural strength results for various bacterial strains.

The results substantiated that the bacteria-incorporated specimens performed well than the control specimens. The *subtilis* specimens yielded a higher increase in flexural strength than the *sphaericus*-incorporated specimens. An increase of about 21% in flexural strength has been reported with *subtilis* and about 17% with *sphaericus*. Jena *et al.* [24] reported an increase of about 29% in the flexural strength of cement concrete specimens with the *subtilis* strain. The reason for the increase in the hardened property is the microbial activity of the bacterial strains inside the concrete.

11.5.4 Water Absorption

GPC with different bacterial strains cast in cubical specimens was subjected to the water absorption test, and the results are shown in Fig. 11.4.

The analysis of water absorption by concrete specimens revealed significant reductions with the addition of bacterial additives. Comparing specimens without bacteria to those containing the *subtilis* strain, a notable percentage decrease of approximately

23.26% in water absorption was observed. Similarly, incorporating the *sphaericus* strain of bacteria resulted in a decrease of around 18.60% in water absorption compared to the control specimens. These findings underscore the effectiveness of both the *subtilis* and *sphaericus* strains in mitigating water ingress into concrete structures, indicating improved durability and resistance to environmental deterioration. The observed reductions in water absorption suggest promising applications for bacterial additives in enhancing the performance and longevity of concrete infrastructures, particularly in environments prone to moisture-related degradation.

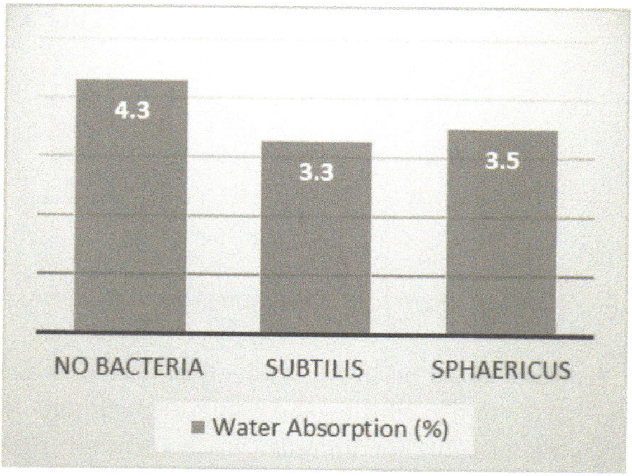

Figure 11.4 Water absorption results for various bacterial strains.

This research investigated the effects of incorporating both *B. subtilis* and *B. sphaericus* bacteria on the strength of concrete specimens. Compressive, splitting tensile, and flexural strengths were evaluated at 7 and 28 days after casting. The findings revealed notable enhancements in strength characteristics with the addition of both bacterial strains. Specifically, at 7 days, the addition of the *subtilis* strain exhibited an 11.23% increase in compressive strength, a 14.81% increase in splitting tensile strength, and a 20.51% increase in flexural strength compared to specimens without bacteria. Similarly, the addition of the *sphaericus* strain demonstrated an 8.33% increase in compressive strength, a 14.81% increase in splitting tensile strength, and a 15.38% increase in flexural strength compared to specimens without bacteria at the same age. At 28 days, the addition of *subtilis* displayed an 11.88% increase in compressive strength, a 13.89% increase in splitting tensile strength, and a 13.73% increase in flexural strength compared to specimens without bacteria, while the addition of *sphaericus* exhibited a 10.14% increase in compressive strength, an 11.11% increase in splitting tensile strength, and a 9.80% increase in flexural strength compared to specimens without bacteria. These findings underscore the potential of both the *subtilis* and *sphaericus* strains as additives for enhancing the mechanical properties of concrete, offering promising avenues for the development of more durable and resilient concrete structures in various engineering applications.

11.5.5 Self-Healing Property

The self-healing property of GPC was inspected by introducing intentional cracks using the compression testing machine by applying a load of 3 MPa/min. The loading was continued until the concrete showed a crack of 1 mm breadth. The breadth of the crack changed from 500 μm to 1 mm. The cracks were sprinkled with water for initiating the bacterial action. Intermittent photographs were taken to record the healing of concrete. The healing behavior of the *subtilis*-incorporated specimen is shown in Fig. 11.5 and that of *sphaericus* is shown in Fig. 11.6.

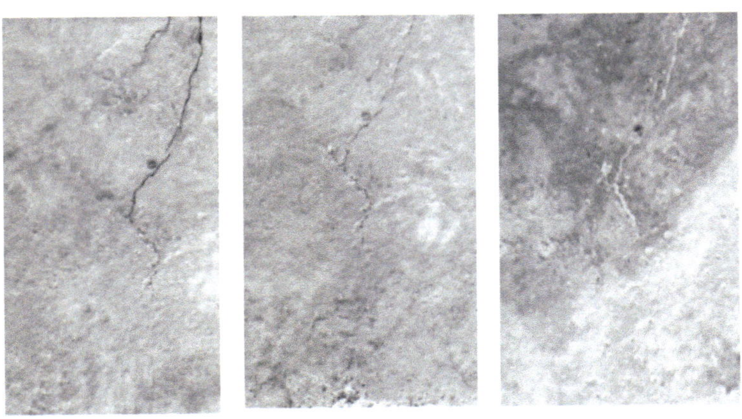

(a)5D (b)14D (c)28D

Figure 11.5 Sealing of cracks using *B. subtilis*.

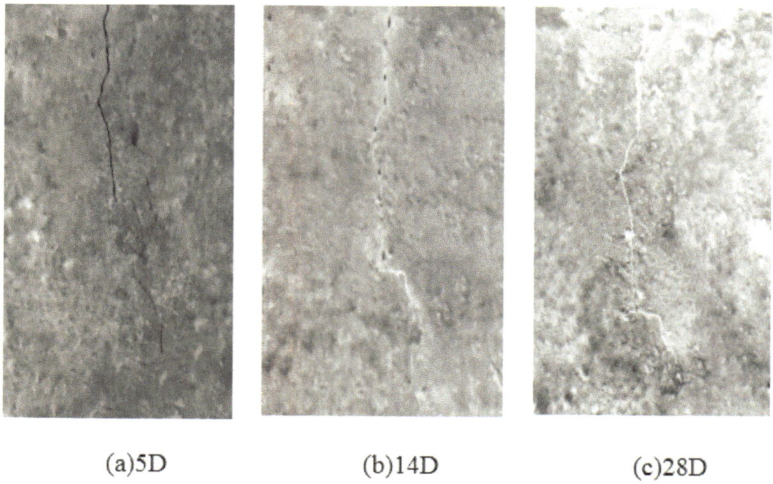

Figure 11.6 Sealing of cracks using *B. sphaericus*.

From Figs. 11.5 and 11.6, it is explicit that after 5 days, the ends of the cracks started to heal with the precipitation of bacteria. On the 14th day, almost half of the cracks were sealed, and on the 28th day, almost all of the cracks were sealed. The application of water inside the cracks initiated the bacterial action and precipitated the calcium carbonate. The white precipitation by the bacteria is calcite and is responsible for the healing action. *B. subtilis* can precipitate calcite even in high alkaline environments. Vijay *et al.* [25] have claimed that the bacteria return to the hibernation stage and wait dormant till the appearance of the next crack and the filling of water.

11.6 Conclusion

The following summary can be drawn from the above results and discussions:

- With the incorporation of both the bacterial strains, an increase of about 11% was reported in the compressive strength of GPC.
- STS witnessed an increase of about 15% with the accumulation of the *subtilis* bacterial strain and 11% with the accumulation of the *sphaericus* bacterial strain.

- The *B. subtilis* and *B. sphaericus* bacterial strains increased the flexural strength of GPC by about 21% and 17%, respectively.
- The *B. subtilis* strain has been shown to be more effective than the *B. sphaericus* strain in improving the properties of GPC.
- Photographs taken at regular intervals reported the effective healing of cracks with the utilization of both the strains of bacteria.

This chapter unleashes the hefty potential of bacterial GPC and paves the way for the development of sustainable construction materials. Further research can focus on evaluating the combined performance of both the bacterial strains.

References

1. Davidovits, J. 1991. Geopolymers. *Journal of Thermal Analysis*, **37**(8): 1633–1656. https://doi.org/10.1007/bf01912193.
2. Chithambar, G. A., and M. Muthukannan. 2018. A review of recent developments in geopolymer concrete. *International Journal of Engineering & Technology*, **7**(4.5): 696. https://doi.org/10.14419/ijet.v7i4.5.25061.
3. Yadav, B. M., Yeswanth, K. K., Bhavani. 2018. Durability characteristics of hybrid fiber reinforced geopolymer concrete with M-sand. *SSRG International Journal of Civil Engineering*, **5**(11): 1–5. https://doi.org/10.14445/23488352/IJCE-V5I11P101
4. Ganesh, C., M. Muthukannan, A. S. Kumar, and K. Arunkumar. 2021. Influence of bacterial strain combination in hybrid fiber reinforced geopolymer concrete subjected to heavy and very heavy traffic condition. *Journal of Advanced Concrete Technology*, **19**(4): 359–369. https://doi.org/10.3151/jact.19.359.
5. Chithambar, G. A., K. Mukilan, B. P. V. Srikar, L. V. S. Teja, K. S. V. Prasad, A. A. Kumar, and R. P. Sharath. 2022. Effect of flyash-rice husk ash blend in the energy efficient geopolymer tiles using industrial wastes. *Materials Science Forum*, **1048**: 40311. https://doi.org/10.4028/www.scientific.net/msf.1048.403.
6. Ganesh, A. C., and M. Muthukannan. 2019. "Effect of elevated temperature over geopolymer concrete. *International Journal of Engineering and Advanced Technology*, **9**(1S3): 450–453. https://doi.org/10.35940/ijeat.a1025.1291s419.

7. Arunachalam, S. K., M. Muthiah, K. D. Rangaswamy, A. Kadarkarai, and C. G. Arunasankar. 2021. Improving the structural performance of reinforced geopolymer concrete incorporated with hazardous heavy metal waste ash. *World Journal of Engineering*, **19**(6): 808–821. https://doi.org/10.1108/wje-01-2021-0055.

8. Ganesh, A C., and M. Muthukannan. 2019. Investigation on the glass fiber reinforced geopolymer concrete made of M-sand. *Journal of Materials and Engineering Structures*, **6**: 501–512.

9. Arunkumar, K., Muthukannan Muthiah, Suresh Kumar, A., Chithambar Ganesh, A., Kanniga Devi, R., 2021. Invention of sustainable geopolymer concrete made with low calcium waste wood ash. *World Journal of Engineering*, **19**(6), 843–853. https://doi.org/10.1108/WJE-03-2021-0162

10. Suresh, K., M. Muthukannan, A. K. Kumar, G. A. Chithambar, and D. R. Kanniga. 2022. Influence of incinerated biomedical waste ash and waste glass powder on the mechanical and flexural properties of reinforced geopolymer concrete. *Australian Journal of Structural Engineering*, **23**(3): 254–268. https://doi.org/10.1080/13287982.2022.2044613.

11. Kumar, S., Arunachalam, M. Muthukannan, A. Kadarkarai, C. G. Arunasankar, and D. R. Kanniga. 2021. Utilisation of waste glass powder to improve the performance of hazardous incinerated biomedical waste ash geopolymer concrete. *Innovative Infrastructure Solutions*, **7**(1). https://doi.org/10.1007/s41062-021-00694-8.

12. Kumar, S., J. Mishra, S. Kumar, S. Mohammed, A. Patel, S. Kumar, and U. Behera. 2020. Materials today: Proceedings characterization and utilization of rice husk ash (RHA) in fly ash – blast furnace slag based geopolymer concrete for sustainable future. *Materials Today Proceedings*, **33**(8): 5163–5164. DOI:10.1016/j.matpr.2020.02.870.

13. Detphan, S., and P. Chindaprasirt. 2009. Preparation of fly ash and rice husk ash geopolymer. *International Journal of Minerals, Metallurgy and Materials*, **16**(6): 720–726. https://doi.org/10.1016/S1674-4799(10)60019-2.

14. Hwang, C.-L., and T.-P. Huynh. 2015. Effect of alkali-activator and rice husk ash content on strength development of fly ash and residual rice husk ash-based geopolymers. *Construction and Building Materials*, **101**: 1–9. https://doi.org/10.1016/j.conbuildmat.2015.10.025.

15. Liang, G., H. Zhu, Z. Zhang, Q. Wu, and J. Du. 2019. Investigation of the waterproof property of alkali-activated metakaolin geopolymer added with rice husk ash. *Journal of Cleaner Production*, **230**: 603–612. https://doi.org/10.1016/j.jclepro.2019.05.111.

16. Tanyildizi, H., M. Ziada, M. Uysal, N. Doğruöz Güngör, and A. Coşkun. 2022. Comparison of bacteria-based self-healing methods in metakaolin geopolymer mortars. *Case Studies in Construction Materials,* **16**: e00895. https://doi.org/10.1016/j.cscm.2022.e00895.
17. Sankarapandian, K., D. L. Venkatesh Babu, and G. Prince Arulraj. 2021. Influence of *Bacillus licheniformis* on the strength, durability and micro-structure of concrete. *Research Square.* https://doi.org/10.21203/rs.3.rs-970612/v1.
18. Krishnapriya, S., D. L. Venkatesh Babu, and G. Prince Arulraj. 2015. Isolation and identification of bacteria to improve the strength of concrete. *Microbiological Research,* **174**: 48–55. https://doi.org/10.1016/j.micres.2015.03.009.
19. Anuradha, R., V. Sreevidya, R. Venkatasubramani, and B. Vijaya Rangan. 2012. Modified guidelines for geopolymer concrete mix design using Indian standard. *Asian Journal of Civil Engineering (Building and Housing),* **13**(3): 357–368.
20. Ashwij, L., Nikhil T R, B. C. Shanthappa, M. R. Prakash, and S. M. Gopinath. 2019. A study on strength, workability and fire resistance properties of bacteria induced concrete. *International Journal of Civil Engineering,* **6**(9): 12–15, https://doi.org/10.14445/23488352/ijce-v6i9p103.
21. Mutitu, D. K., J. Muthengia Wachira, R. Mwirichia, J. Karanja Thiong'o, O. M. Munyao, and G. Muriithi. 2019. Influence of *Lysinibacillus sphaericus* on compressive strength and water sorptivity in microbial cement mortar. *Heliyon,* **5**(11): e02881. https://doi.org/10.1016/j.heliyon.2019.e02881.
22. Shanmuga, P. T., N. Ramesh, A. Agarwal, S. Bhusnur, and K. Chaudhary. 2019. Strength and durability characteristics of concrete made by micronized biomass silica and bacteria-*Bacillus sphaericus. Construction and Building Materials,* **226**: 827–838. https://doi.org/10.1016/j.conbuildmat.2019.07.172.
23. Meera, C. M., and V. Subha. 2016. Strength and durability assessment of bacteria based self-healing concrete. *IOSR Journal of Mechanical and Civil Engineering,* (2016): 8-1–7.
24. Jena, S., B. Basa, K. C. Panda, and N. K. Sahoo. 2020. Impact of *Bacillus subtilis* bacterium on the properties of concrete. *Materials Today: Proceedings,* **32**: 651–656. https://doi.org/10.1016/j.matpr.2020.03.129.
25. Vijay, K., M. Murmu, and S. V. Deo. 2017. Bacteria based self-healing concrete: A review. *Construction and Building Materials,* **152**: 1008–1014. https://doi.org/10.1016/j.conbuildmat.2017.07.040.

Chapter 12

Geopolymer Bricks Using Industrial Wastes to Reduce CO_2 Footprint

A. Chithambar Ganesh,[a] R. Mohana,[b] M. Vinod Kumar,[a] V. Prem Kumar,[c] and Mehmet Serkan Kirgiz[d]
[a]*Vel Tech Rangarajan Dr. Sagunthala R&D Institute of Science and Technology, Chennai, India*
[b]*Mepco Schlenk Engineering College, Sivakasi, India*
[c]*Sree Vidyanikethan Engineering College, Tirupati, India*
[d]*Academia of ACMCEN, Beaverton, OR 97075, USA*
nakres42@yahoo.com

The most commonly and widely used masonry units are bricks. Brickmaking is an energy-intensive operation that contributes significantly to air pollution through the liberation of CO_2. The main objective of this chapter is to develop geopolymer bricks based on industrial waste under ambient conditions. The binding material used here is ground granulated blast furnace slag (GGBS) for the mixer of sodium hydroxide (SH) and sodium silicate (SS) solution as the alkaline enhancer. Iron slag is used as the fine aggregate. The key features that influence the properties of geopolymer bricks—proportion of GGBS and iron slag, ratio of SH(NaOH) and

Handbook of Innovative Adhesive Technology: Biodegradable and Cost-Effective Binder for Built Environment and Engineering Sciences
Edited by Mehmet Serkan Kirgiz

ISBN 978-981-5129-51-9 (Hardcover), 978-1-003-59892-3 (eBook)
www.jennystanford.com

SS(Na_2SiO_3), and the molarity of NaOH solution—are optimized in this chapter. Brick tests are in accordance with IS 3495 (Part-2): 1992. This chapter aims to develop geopolymer bricks with qualities comparable to Class A first-class bricks with the least strain on the environment. This chapter would be beneficial in the arena of bricks and building materials.

12.1 Introduction

Since 14,000 BC, bricks have been used as a building material and are still widely utilized in construction today. Bricks are predominantly made from clay. Clay bricks have high strength and durability and are, therefore, irreplaceable. Traditional clay brickmaking methods are comparable in many parts of the world. The fundamentals of conventional clay brick production have not altered throughout time [1]. They differ only in the building of the brick kiln. Clay brick manufacturing entails baking bricks at high temperatures of around 1400°C inside a kiln [2]. The process involves the use of a significant amount of fuel, such as wood, coal, or biomass, which produces substantial air pollution. About 24 million tonnes of coal are burned every year. Brick kilns pose a major threat to the health of local residents due to the release of harmful compounds such as ammonia, CO, and CO_2 into the environment [3]. Hence, for a sustainable development, an energy-efficient method of making bricks must be developed.

Bricks can be manufactured at temperatures as low as 600°C using river sand and sludge [4]. River sand is a nonrenewable resource, and it leaves strain over the environment when it is mined. Its depletion must be managed to ensure long-term development [5]. Developments in the building sector have created increased demands for river sand. Hence, other alternative materials might be preferred, which are unutilized and available in abundance. Currently, mining has become the most important economic activity in most developing countries [6].

The expansion of iron and steel industries has increased the production of slag, which has a notable environmental risk. A large quantity of by-products are generated, and discarding them has turned to a major challenge. A majority of these wastes are landfilled, which is incompatible with sustainable growth. Owing to a shortage of competent labor and an inadequate budget for trash removal, open dumpsites are prevalent in developing countries such as India. Improper handling of waste, such as transportation, disposal, or storage, can cause health and environmental hazards [7].

The production of geopolymer bricks has gained popularity recently due to its environmentally beneficial characteristics. Geopolymer technology is believed to be environmentally benign since it helps to reduce carbon dioxide emissions from cement manufacture while also solving the issue of fly ash and GGBS disposal [8]. Geopolymer technology can be extended to brick manufacturing using materials such as fly ash [9–11], GGBS [12–14], rice husk ash [15], wood ash [16], glass powder [17], and biomedical waste ash [18]. Geopolymer brick qualities vary greatly based on the amounts of source materials used, the ratio of Na_2SiO_3 and NaOH alkaline solutions employed, and the concentration of NaOH solution [19].

This chapter deals with a set of experiments to determine the influence of the different factors that affect the sturdiness and water absorption properties of geopolymer bricks. The importance of this chapter lies in the utilization of iron slag as the fine aggregate for GGBS-based geopolymer mortar bricks. The only limitation in the acceptance of geopolymer building materials as a green alternative is the requirement of an alkaline activator solution for their synthesis. The major challenge is reducing the requirement and concentration of the solution for making geopolymer bricks.

12.2 Materials and Methodology

Materials required for making geopolymer bricks include GGBS as the binder source material, iron slag as the fine aggregate, and a concoction of SH(NaOH) and SS(Na_2SiO_3) solution as the alkaline enhancer. The specific gravities of GGBS and iron slag are found to be 2.9 and 2.62, respectively. The specific gravities of NaOH and Na_2SiO_3 are 1.47 and 1.6, respectively. The materials are proportioned based on fact findings. GGBS and iron slag are initially mixed dry in a dry mixer, and then the activator solution is added. For the next 5 min, the mixing is continued and is casted in molds of dimension 190 × 90 × 90 mm.

The investigation is divided into three phases because different elements that affect the qualities of geopolymer bricks had to be optimized. In the first phase, the raw ingredients needed for developing geopolymer bricks, such as GGBS and iron slag, are varied as 70/30, 60/40, 50/50, 40/60, and 30/70 and optimized. The optimal percentage is chosen, and the sodium silicate/sodium hydroxide solution ratio to be utilized is then varied as 2.5, 2, 1.5, and 1 and optimized. Finally, the molarity of the sodium hydroxide solution is varied as 8, 10, 12, 14, and 16 M and optimized.

12.3 Results and Discussions

12.3.1 Effect of GGBS–Iron Slag Content

In this phase, the iron slag/GGBS ratio is varied as 70/30, 60/40, 50/50, 40/60, 30/70, and 20/80. The source materials GGBS and iron slag are optimized for compressive strength, efflorescence, and water absorption test as per IS 1077:1992. The values are listed in Table 12.1.

Table 12.1 Mix proportions with variations of GGBS:iron slag and test values

Specimen Type	Iron Slag (%)	GGBS (%)	SS:SH	Molarity of SH	CS (MPa)	WS (%)	Efflorescence	
							(%)	Category
GI1	70	30	1	8	8.6	15.2	45	Moderate
GI2	60	40	1	8	10.2	10.6	29	Moderate
GI3	50	50	1	8	15.3	8.2	15	Moderate
GI4	40	60	1	8	16.5	6.3	14	Moderate
GI5	30	70	1	8	22.4	5.6	8	Slight
GI6	20	80	1	8	26.4	5.3	5	Slight

GI6 reveals both maximum compressive strength and minimum water absorption capacity. GI6 also exhibits the best efflorescence under slight category, which is less than 10%. From Table 12.1, it can be observed that the compressive strength gradually rises when the GGBS/iron slag ratio rises. This is because of the presence of more calcium (Ca) content in the GGBS, which is the base material [9]. There is a consequent reduction in the absorption of water capacity with the surge in GGBS content in the mortar. Figure 12.1 depicts the effect of the iron slag/GGBS ratio on compressive strength and water absorption.

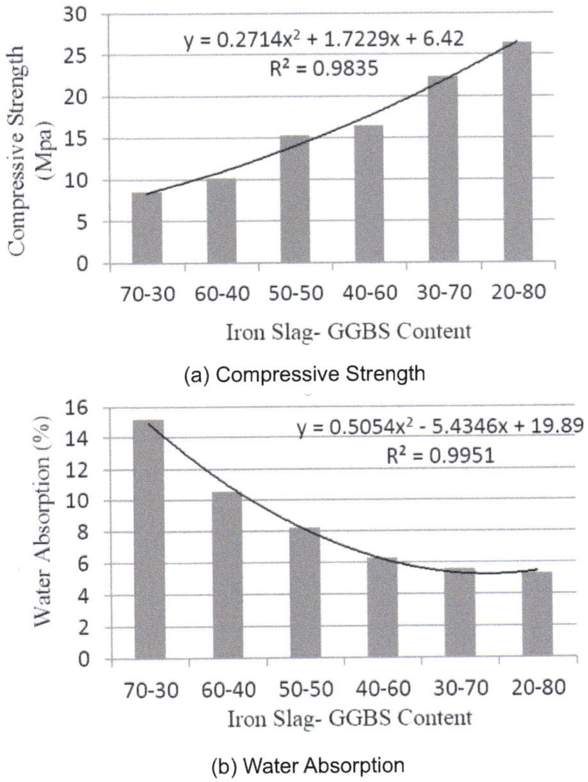

Figure 12.1 Effect of iron slag and GGBS ratio.

When the water absorption capacity is low, it tends to show better morphology in the matrix. The results of compressive strength are indirectly proportional to the water absorption results. In this phase, specimen GB12 with compressive strength just more than 10 MPa was selected for the investigation, and the utilization of more GGBS would result in the increased consumption of the alkaline activator solution.

The analysis of the data in Table 12.1 reveals a correlation between the increase in GGBS content and the corresponding gain in compressive strength, which is the characteristic strength of the specimens. When the specimens are compared, the percentage gain in compressive strength demonstrates varying trends. For instance, the transition from GI1 to GI2, with a rise in GGBS content from 30% to 40%, results in a percentage gain of approximately 18.60%. This gain notably increases with higher proportions of GGBS, as seen in the shift from GI2 to GI3, where the percentage gain jumps to around 50.00% with a corresponding GGBS increase from 40% to 50%. However, as GGBS content continues to increase, the percentage gain diminishes slightly. For instance, the transition from GI4 to GI5, with an increase in GGBS from 60% to 70%, yields a percentage gain of approximately 35.76%. This trend continues with the transition from GI5 to GI6, indicating a percentage gain of approximately 17.86% with a further increase in GGBS content from 70% to 80%. Overall, the analysis underscores the influence of GGBS content on compressive strength, suggesting a nonlinear relationship where the gains diminish marginally as the GGBS content reaches higher proportions.

Upon analyzing the data, it is evident that there is a notable relationship between the increase in GGBS content, the corresponding rise in compressive strength, and the decrease in water absorption across the specimens. As the GGBS content increases from GI1 to GI6, there is a consistent trend of increased compressive strength, indicating the potential enhancement of the material's structural integrity. This enhancement is particularly pronounced in transitions such as from GI2 to GI3 and from GI4 to GI5, where the increase in compressive strength is substantial, indicating the significant influence of higher GGBS proportions on strength improvement. Concurrently, there is a consistent decrease in water absorption across the specimens, suggesting improved resistance to moisture ingress. This trend underscores the potential benefits of higher GGBS content in reducing the material's permeability and enhancing its durability, which is crucial for applications in construction and infrastructure. However, it is important to note that the magnitude of change in compressive strength and water absorption varies across transitions, indicating potential nonlinear relationships or other influencing factors that warrant further investigation for a comprehensive understanding of the material's behavior.

12.3.2 Effect of Na_2SiO_3 and NaOH

The second phase involves the optimization of the alkaline activator solution for compressive strength, efflorescence, and water absorption test as per IS 1077-1992. The Na_2SiO_3 : NaOH ratio is varied as 2.5, 2, 1.5, and 1. The values are listed in Table 12.2.

Table 12.2 Mix proportions with variation in SS:SH and test values

Specimen Type	Iron Slag (%)	GGBS (%)	SS:SH	Molarity of SH	CS (MPa)	WS (%)	Efflorescence (%)	Efflorescence Category
GI2	60	40	1	8	10.2	10.6	15	Moderate
GI2a	60	40	1.5	8	11.5	8.6	11	Moderate
GI2b	60	40	2	8	13.8	8.2	8	Slight
GI2c	60	40	2.5	8	14.6	7.7	5	Slight

GI2c exhibits the maximum compressive strength of 14.6 MPa with about 43% increase. From Table 12.2, it is evident that with an increase in the ratio of SS:SH, there is a constant increment in the compressive strength and a decrease in the water absorption and efflorescence values. At an increased sodium silicate concentration, the rate of polymerization reaction is enhanced [20]. Figure 12.2 depicts the effect of the SS:SH ratio on compressive strength and water absorption.

As the SS:SH ratio increases from 1 to 2.5, there is a corresponding increase in compressive strength (CS) from 10.2 MPa for GI2 to 14.6 MPa for GI2c. This trend suggests a positive correlation between the SS:SH ratio and compressive strength, indicating that higher ratios contribute to greater strength. Additionally, as the SS:SH ratio increases, there is a noticeable decrease in water absorption (WS), with GI2 recording 10.6% and GI2c demonstrating 7.7%. This inverse relationship between SS:SH ratio and water absorption implies that higher ratios result in reduced permeability, enhancing the material's durability against moisture ingress.

Moreover, the efflorescence percentages decrease with an increasing SS:SH ratio, from 15% for GI2 to 5% for GI2c, indicating reduced salt deposition on the material surface. Overall, these findings suggest that adjusting the SS:SH ratio in the composition can significantly influence the material's compressive strength, water absorption, and efflorescence, thereby offering opportunities for optimizing its performance in various applications.

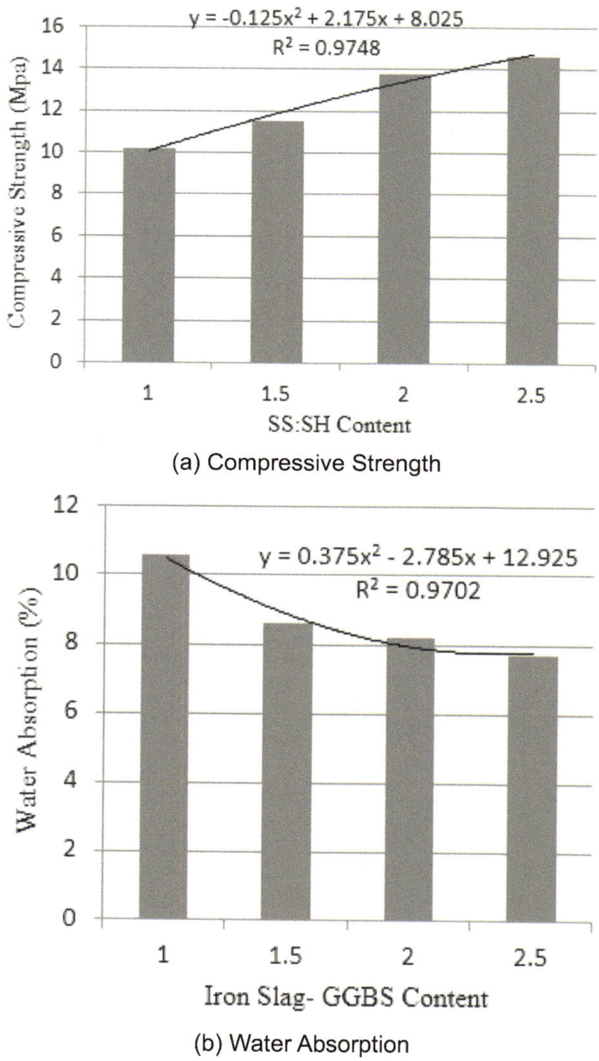

Figure 12.2 Effect of the SS:SH ratio.

The least water absorption capacity is due to the flawless inner structure, which is the result of better geopolymerization with the higher SS:SH ratio. The slight category of the efflorescence value reports the presence of the least amount of salts in the outer surface of geopolymer bricks. The synthesis of sodium silicate solution imparts least strain over the environment when compared to sodium hydroxide solution, and hence a higher SS:SH ratio can be preferred for the next phase.

12.3.3 Effect of Molarity

The molarity of the SH(NaOH) solution is optimized for compressive strength, efflorescence, and water absorption test in accordance with IS 1077-1992. The molarity of the SH solution is varied as 2.5, 2, 1.5, and 1. The values are listed in Table 12.3.

Table 12.3 Mix proportions with variation in molarity and test values

Specimen Type	Iron Slag (%)	GGBS (%)	SS:SH	Molarity of SH	CS (MPa)	WS (%)	Efflorescence (%)	Efflorescence Category
GI2c	60	40	2.5	8	14.6	7.7	5	Slight
GI2c10	60	40	2.5	10	15.1	7.3	-	Nil
GI2c12	60	40	2.5	12	15.9	6.4	-	Nil
GI2c14	60	40	2.5	14	15.3	6.5	-	Nil

Table 12.3 shows specimens designated as GI2c, GI2c10, GI2c12, and GI2c14, each maintaining consistent iron slag (60%) and GGBS (40%) compositions but varying molarities of SH and consequent compressive strength (CS), water absorption (WS), and efflorescence percentages. Notably, as the molarity of SH increases from 8 in GI2c to 14 in GI2c14, there is a corresponding increase in compressive strength from 14.6 MPa in GI2c to 15.3 MPa in GI2c14. This trend suggests a positive correlation between the molarity of SH and compressive strength, implying that higher molarities contribute to enhanced strength properties. Additionally, as the molarity of SH increases, there is a consistent decrease in water absorption, with GI2c recording 7.7% and GI2c14 demonstrating 6.5%.

Figure 12.3 depicts the effect of molarity on compressive strength and water absorption. This indicates that higher molarities of SH result in reduced water absorption, improving the material's resistance to moisture ingress. Notably, all specimens from GI2c10 to GI2c14 exhibit no efflorescence, suggesting minimal salt deposition on the material surface. Overall, these findings underscore the significant impact of adjusting the molarity of SH on the material's compressive strength, water absorption, and efflorescence, highlighting avenues for optimizing its performance in various applications.

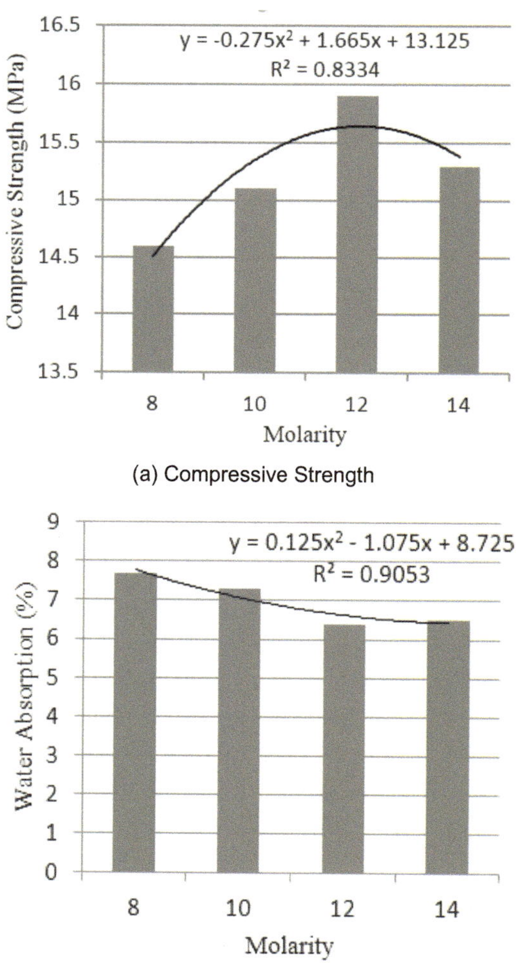

(a) Compressive Strength

(b) Water Absorption

Figure 12.3 Effect of molarity on compressive strength and water absorption.

A decrease in strength is observed beyond 12M as the excess NaOH ions remain inert and delay the polymerization reaction [21]. Also, the other difficulty observed with the solution of higher molarity is the precipitation of salts in the beaker bottom. This is clear evidence of the decrease in strength at a higher molarity beyond 12 M.

12.4 Conclusion

The investigation of the development of geopolymer bricks throws light on ways to reduce carbon footprint and certain other conclusions as follows:

- Geopolymer bricks could be synthesized using 60% iron slag mixed with 40% GGBS to attain a compressive strength at par with first-class bricks.
- A prominent increase in compressive strength of about 43% has been observed with an increase in the ratio of SS(Na_2SiO_3) and SH(NaOH) solution of 2.5:1.
- A considerable reduction in water absorption capacity and efflorescence with an increase in the ratio of SS(Na_2SiO_3) and SH(NaOH) solution has been observed.
- Sturdiness increases with an increase in molarity till 12M and after that it decreases.
- Water absorption capacity and efflorescence decrease till 12M and then increase.

It can be understood that geopolymer bricks so reported could be prepared at ambient curing conditions with minimum enhancers, and the minimum concentration of sodium hydroxide (NaOH) solution, G12c, with the minimum molarity of 8M exhibits strength equivalent to first-class bricks.

References

1. Dalkilic, N., and A. Nabikoğlu. Traditional manufacturing of clay brick used in the historical buildings of Diyarbakir (Turkey). *Frontiers of Architectural Research,* **6**, 3 (2017): 346–359. https://doi.org/10.1016/j.foar.2017.06.003.

2. Sorte, A. M., A. N. Burile, A. Chaudhari, and A. G. M. Haldar. Utilisation of agro waste in the development of fired clay bricks: A review. *International Journal of Environment and Waste Management,* **26**, 4 (2020): 531. https://doi.org/10.1504/ijewm.2020.110400.
3. Heindl, R. A., and W. L. Pendergast. Progress report on investigation of fireclay brick and the clays used in their preparation. *Journal of the American Ceramic Society,* **12**, 10 (1929): 640–675. https://doi.org/10.1111/j.1151-2916.1929.tb18006.x.
4. Lissy, P., C. Peter, K. Mohan, S. Greens, and S. M. George. Energy efficient production of clay bricks using industrial waste. *Heliyon,* **4**, 10 (2018): e00891. https://doi.org/10.1016/j.heliyon.2018.e00891.
5. Padmalal, D., K. Maya, S. Sreebha, and R. Sreeja. Environmental effects of river sand mining: A case from the river catchments of Vembanad Lake, Southwest Coast of India. *Environmental Geology,* **54**, 4 (2007): 879–889. https://doi.org/10.1007/s00254-007-0870-z.
6. Walsh, B., S. Van Der Plank, and P. Behrens. The effect of community consultation on perceptions of a proposed mine: A case study from Southeast Australia. *Resources Policy,* **51** (2017): 163–171. https://doi.org/10.1016/j.resourpol.2016.12.006.
7. Misra, V., and S. D. Pandey. Hazardous waste, impact on health and environment for development of better waste management strategies in future in India. *Environment International,* **31**, 3 (2005): 417–431. https://doi.org/10.1016/j.envint.2004.08.005.
8. Van Deventer, J. S. J., Provis, J. L., and Duxson, P. Technical and commercial progress in the adoption of geopolymer cement. *Minerals Engineering,* **29** (2012): 89–104. doi:10.1016/j.mineng.2011.09.009.
9. Ganesh, A. C., M. Kumar, R. G. Devi, P. Srikar, S. V. Prasad, M. Kumar, and R. P. Sarath. Pervious geopolymer concrete under ambient curing. *Materials Today: Proceedings,* **46** (2021): 2737–2741. https://doi.org/10.1016/j.matpr.2021.02.425.
10. Ganesh, A. C., K. Mukilan, B. P. V. Srikar, L. V. S. Teja, K. S. V. Prasad, A. S. Kumar, and R. P. Sharath. Effect of flyash-rice husk ash blend in the energy efficient geopolymer tiles using industrial wastes. *Materials Science Forum,* **1048** (2022): 403–411. https://doi.org/10.4028/www.scientific.net/msf.1048.403.
11. Ganesh, A. C., M. Muthukannan, S. Aakassh, Prasad, and B. Subramanaian. Energy efficient production of geopolymer bricks using industrial waste. *IOP Conference Series,* **872** (2020): 012154. https://doi.org/10.1088/1757-899x/872/1/012154.

12. Arunachalam, S. K., M. Muthukannan, K. Arunkumar, A. C. Ganesh, and R. K. Devi. Utilisation of waste glass powder to improve the performance of hazardous incinerated biomedical waste ash geopolymer concrete. *Innovative Infrastructure Solutions,* **7**, 1 (2022). https://doi.org/10.1007/s41062-021-00694-8.

13. Ganesh, A. C., and M. Muthukannan. Investigation on the glass fiber reinforced geopolymer concrete made of M-sand. *Journal of Materials and Engineering Structure,* **6**, 4 (2019), 501–512.

14. Shaikh, F. U. A., and Haque, S. Alkali Activated fly ash and slag bricks for masonry construction. *International Journal of Concrete Structures and Materials,* **12**, 1 (2018): 1–13. doi: 10.1186/s40069-018-0272-x.

15. Detphan, S., and P. Chindaprasirt. Preparation of fly ash and rice husk ash geopolymer. *International Journal of Minerals Metallurgy and Materials,* **16**, 6 (2009): 720–726. https://doi.org/10.1016/s1674-4799(10)60019-2.

16. Arunkumar, K., M. Muthukannan, A. S. Kumar, A. C. Ganesh, and R. K. Devi. Production of eco-friendly geopolymer concrete by using waste wood ash for a sustainable environment. *Pollution,* **7**, 4 (2021): 993–1006. https://doi.org/10.22059/poll.2021.320857.1039.

17. Kumar, S. A., M. Muthukannan, K. A. Kumar, A. C. Ganesh, and R. K. Devi. Influence of incinerated biomedical waste ash and waste glass powder on the mechanical and flexural properties of reinforced geopolymer concrete. *Australian Journal of Structural Engineering,* **23**, 3 (2022): 254–268. https://doi.org/10.1080/13287982.2022.2044613.

18. Suresh, K., A. Suresh, M. Muthukannan, R. Gayatri Devi, K. Arunkumar, and A. C Ganesh. Reduction of hazardous incinerated bio-medical waste ash and its environmental strain by utilizing in green concrete. *Water Science and Technology,* **84**, 10–11 (2021): 2780–2792. https://doi.org/10.2166/wst.2021.239.

19. Ganesh, A. C, and M. Muthukannan. A review of recent developments in geopolymer concrete. *International Journal of Engineering & Technology,* **7**, 4.5 (2018): 696. https://doi.org/10.14419/ijet.v7i4.5.25061.

20. Revathi, V., J. Thaarrini, and M. Venkob Rao. A prospective study on alkali activated bottom Ash-GGBS blend in paver blocks. *Journal of Civil and Environmental Engineering,* **8**, 3 (2014): 291–298. https://publications.waset.org/9997832/pdf.

21. Deb, P., P. Nath, and P. K. Sarker. Strength and permeation properties of slag blended fly ash based geopolymer concrete. *Advanced Materials Research,* **651** (2013): 168–173. https://doi.org/10.4028/www.scientific.net/amr.651.168.

Chapter 13

Natural Rubber Latex–Substituted Bitumen Binder for Road Construction: A Review

Trinity A. Tagbor,[a] Mehmet Serkan Kirgiz,[b] Theresah Osei,[c] Latifatu Mohammed,[a] Ammishaddai Ofori-Nyarko,[a] Festus Mensah,[c] and Samuel Obiri[c]

[a]*CSIR-Institute of Industrial Research, P.O. Box LG 57, Legon-Accra, Ghana*
[b]*Academia of ACMCEN, Beaverton, OR 97075, USA*
[c]*CSIR-Institute of Building and Road, P.O. Box UP 40, Kumasi, Ghana*
nakres42@yahoo.com

Natural rubber (NR) in bituminous mixes is a great innovation in the production of sustainable and quality binders for mostly road and pavement construction. This chapter reviews some studies on the feasibility of NR in bituminous mixes for road and pavement construction. The nature, properties, and uses of NR, the modification of bitumen using NR, and the effect of NR on bituminous mixes are reviewed and discussed. The review has shown that NR can be used

Handbook of Innovative Adhesive Technology: Biodegradable and Cost-Effective Binder for Built Environment and Engineering Sciences
Edited by Mehmet Serkan Kirgiz

ISBN 978-981-5129-51-9 (Hardcover), 978-1-003-59892-3 (eBook)
www.jennystanford.com

for sustainable construction of pavements and roads. The optimum proportion of NR was reported to be 4–8% of the bituminous mixes from various studies. The increase in viscoelasticity of the resulting binder may help improve the service life of pavements while reducing maintenance costs. However, the limited number of field studies on the performance of NR-modified bitumen suggests the need for further research in this regard to determine the cost-effectiveness of its use as a binder for commercial purposes. Tropical countries, which are also producers of NR and yet rely on imported bitumen for the construction of roads, could boost their domestic consumption and make some savings by incorporating NR in the bituminous mixes for the construction of asphaltic roads.

13.1 Introduction

The demand for nonrenewable resources has increased in recent times amid increasing concerns about the depletion of these resources. Thus, bio-friendly and renewable materials, including plants, waste/residues, or organic matter from various sectors and industries, are increasingly being used as eco-friendly substitutes for nonrenewable materials all over the world. The past decades have, therefore, seen an increase in the demand for eco-friendly materials and green products [1]. NR is a biopolymer material extracted from trees that can be readily used for various applications. NR is an environmentally friendly material, and due to the good adhesive properties it possesses, it is mostly used as a binder for engineering applications. It is used as sealants and binders and for the manufacturing of products such as tires and gloves. The demand for the use of NR either as a source material for production or as a material to aid processing keeps increasing globally. Most roads are paved with asphalt mixtures because they are more economical and flexible than concrete [2]. Even though the proportion of bitumen in asphalt mix is very little, it significantly affects the performance and durability of roads and pavements [3]. Asphalt pavements tend to crack as a result of temperature fluctuations. This has necessitated the search for materials that could enhance the crack resistance property of asphalt binders. In addition to this, the load on pavements and roads in recent years has increased significantly due to the increased number of vehicles on the roads. Thus, there is a push all over the world to find materials that will enhance the durability of roads and pavements.

Moreover, polymer-modified bitumen (PMB) is a technology that has been studied and is used all around the world to increase the service life of pavements and roads. Studies on the use of NR to modify bitumen in hot asphalt mixes started a long time ago [4]. NR is used to hold asphalt together for mostly pavements and roads and also to hold aggregates together in cement-based materials, and this improves several mechanical and physical properties of the resulting products. Ghana, a developing tropical country like many others, is an NR-producing country but relies on imported bitumen for the construction of its asphalt roads. This is associated with high construction and maintenance costs, especially in the wake of the increasing foreign exchange rate and also the early and frequent deterioration of these roads. The use of PMB has recently been introduced into the country, but the polymer being used in most cases is synthetic and imported. Though there are various experimental studies on the use of other polymers, such as waste plastics [5], practical field applications using these are sparse. By reviewing the literature on the use of NR in bituminous mixes for asphalt roads, various studies have been conducted and their findings are highlighted with the view of informing future research to promote the use of NR in bituminous mixes for sustainable road construction.

13.2 Materials and Methodology

Google Scholar, Research Gate, Scielo, and Science Direct were the main databases used for the articles that were reviewed. The review process started with the search for articles on NR and then on NR in bituminous mixes. Keywords for the initial search were "natural rubber latex," "polymer-modified bitumen," "natural rubber in bituminous mixes," among others. A total of about 164 articles were sampled for review. The sampled papers were sorted by reading through the titles, abstracts, and conclusions. All immaterial articles and papers that were difficult to access were removed, which brought the number of articles to be reviewed to 51. Google Scholar was used to refine the remaining articles by using keywords such as "natural rubber latex" and "natural rubber and bitumen." The subsequent sections present information on natural rubber, its uses, and physical and chemical properties; studies on bitumen, its properties and applications; use of natural rubber in bituminous mixes and its advantages; some challenges; and conclusion and recommendations.

13.3 Natural Rubber

NR primarily comprises polyisoprene with remarkable and desirable properties that make it useful for many engineering applications and for the manufacture of products such as hoses, tires, tubes, belts as well as medical products such as gloves. Plants make several cis-polyisoprenes; however, only those that are a minimum 100 isopentenyl units long are considered rubber [6]. Rubber, as one of the most important materials produced by plants, is used for thousands of products worldwide [7]. Rubber-bearing plants are very important for latex allergy products and solid rubber [8]. Rubber produced naturally from these plants are of better quality than those produced synthetically [9]. The molecular structure of NR gives it unique physical properties and make it suitable for many applications [10]. NR is harvested from many rubber-producing plant species, but the rubber tree *Hevea brasiliensis*, which originates from Brazil, has significant commercial value, producing about 99% of all NR used globally [6, 11]. Hevea rubber trees produce latex made up of about 35% rubber [12]. When tapped from the tree, latex is collected in a cup in two ways. First as latex, and for this, ammonia is added to preserve the liquid form. The other procedure is to let the latex coagulate in cups, or formic acid is added to rapidly change latex into lumps [13].

13.3.1 Use of Natural Rubber

NR is used to make a variation of polymer products such as tires, padding of automobile parts, rubber gloves, nozzles, erasers, ducting products, and anti-vibrating and soundproofing materials. NR makes up about 30% of the weight of a tire [14]. It is used for medical products such as latex gloves [9]. There are also some pressure-sensitive adhesives that are made of NR [15]. NR is used as a binder for several applications because it has good adhesive properties.

In the construction of pavements, polymer-modified binders made from NR are added to asphalt to increase the service life of pavements [16]. Escobar and Dos Santos [17] used NR as a component of an eco-friendly binder for ceramic injection molding. Rakesh *et al.* [18] used concentrated NR latex as a binder to cast aluminum powder into tape.

A study by Oluwole and Avwerosuoghene [11], to find the potentials that exist in biodegradable materials and waste products for the development of strong and low-cost engineering materials for ceiling and/or partitioning boards, concluded that NR as a binder imparted better water-repellent or water-resistant property. According to Oluwole and Avwerosuoghene [11], NR-bonded composites are the best material when considering specific strength properties because of NR's low density compared to starch.

13.3.2 Physical Properties of Natural Rubber

NR has similar abilities as most polymeric materials. It is versatile with great tensile and tear strengths [19]. It has good abrasion resistance and slow crack growth [20]. It has high strength-to-weight ratio and good resilience and easily adheres with other materials [21]. It has a melting temperature of about 200°C [22]. NR has special properties that rely on the presence of non-rubber materials. Naturally occurring cross-linking in NR plays a principal role in the mechanical property of the rubber, such as green strength [23]. In addition to all these good qualities, NR has poor oil and flame resistance along with poor aging property. These undesirable attributes may be due to the nonpolar nature of NR and high unsaturation of the molecular chain [24].

13.3.3 Chemical Properties of Natural Rubber

Infrared, NMR, and X-ray studies have shown that NR is made up of cis-polyisoprene with the chemical formula of $(C_5H_8)n$ and a very high molecular weight of about one million g/mol. NR is resistant to most chemical attacks. It has good hydrophobic property [25]. Polyisoprene in NR can be made synthetically due to the double bonds present in each repeated unit, making it liable for vulcanization [21]. NR is significantly affected by non-rubber ingredients such as phospholipids, proteins, carbohydrates, and inorganic salts [26] present in it. NR is usually synthesized by polymerizing isoprene units with a Ti/Al catalyst to produce units in the cis-1,4 configuration up to 98% [12]. Synthetic polyisoprene shows similar properties to NR and is used as an alternative for NR in some applications. It is an environmentally friendly material because of its inert characteristics. It is very soluble in hydrocarbon solvents such as petrol and kerosene [27]. NR has good antiviral permeation and is biodegradable [28].

13.3.4 Natural Rubber as a Binder

NR is used as a binder for various applications. It exhibits good adhesive properties due to its elastomeric nature. Several experiments have been conducted to explore the use of NR as a binder. It has been used as a binder for the silicon anode in lithium batteries to enhance cyclic stability and capacitive performance [29]. NR is used as a binder in ceramic injection molding for several fabrication processes of complex components [17]. NR as a binder for tape-casting alumina was studied by Rakesh *et al.* as an eco-friendly binder in the fabrication process [18]. NR has been used in bituminous mixes mostly for asphalt modification. NR-modified asphalt has high elasticity, aging resistance, stability, tear strength, and fatigue resistance among other properties [21]. The characteristic change in bitumen due to NR occurs when the rubber dissolves into some component fraction of the bitumen and forms an interconnecting matrix of polymer within the bitumen [24]. In extremely cold conditions, rubber in asphalt for pavements acts as an elastic band that dissipates stresses accumulated in asphalt due to the temperature [21].

13.4 Bitumen

Bitumen is a viscoelastic material made up of naturally occurring hydrocarbons and residue of petroleum distillation. It contains hydrocarbons, small amounts of some heteroatoms, and traces of some metals [30]. The detailed components of bitumen depend on the source and manufacturing process [31]. It is an involatile material obtained from crude oil and used in the preparation of asphalt [32]. Bitumen is very sensitive to fluctuating seasons and temperature variations [33]. Bitumen is used in the construction industry because of its waterproofing, sealing, and thermoplastic properties [34]. It is mostly used as a binder in asphalt and also for pavement and roofing applications and in industrial products. The use of bitumen as a construction material started as far back as the Middle Ages. Currently most of the bitumen produced is used in paving as a binder for aggregates to form what is known as asphalt mixes [35]. Bitumen is highly sensitive and affected by sunlight as well as changes in temperature, moisture, and oxygen. This leads to changes in the physical characteristics of bitumen, which then affects the mechanical properties of asphalt mixtures. One of the properties of bitumen is aging, which in turn alters the mechanical properties of asphalt mixtures, leading to bitumen not being able to withstand long-term heavy traffic and adverse weather conditions. These shortfalls necessitated research into bitumen modification to enhance the physical and mechanical properties of bitumen for durable and sustainable construction works [36].

13.5 Natural Rubber as a Binder in Bituminous Mixes

Climate conditions such as rain and heat affect asphalt binders more than they do for aggregates. Hence, the durability of pavements is significantly dependent on the binder [37]. Bitumen needs to be modified because it undergoes several kinds of failure during service time. Polymer, therefore, is used to alter the properties of bitumen for pavements and roads and make them more stable and tougher at high temperatures and better flexibility at low temperatures [38]. Although there are other modifiers for bituminous mixes, polymer modifiers (including NR) prove to be much economical and environmentally friendly and are known to be one of the most common binders used globally. NR is sometimes used to modify bitumen for road construction in either a latex or a coagulated form [39]. The use of NR in bitumen is a practice that started long before crumb rubber [3]. NR does not chemically bond with bitumen or change the chemical nature of bitumen when they are mixed [24]. When NR is mixed with bitumen at high temperatures for a period of time, it interacts with NR, resulting in the bitumen being absorbed by the polymer and causing a swelling [40]. This swelling is the first stage of a two-step process of forming a new substance. The first step is the swelling of the bitumen, and the latter step is a chemical degradation process, which is a scission of cross-linked and main chain bonds [41].

NR in bituminous mixes enhances the creep resistance, thermal sensitivity, tensile strength, rutting resistance, among others. Experiments on bitumen modified with NR saw a significant decrease in penetration for very high NR addition [41]. Wen *et al.* [42] reported an enhancement in bitumen viscosity when modified with NR. These changes were attributed to the interaction between NR molecules and bitumen, which makes it compatible and viscous [40]. The blending of NR in bitumen is dependent on how well NR disperses in bitumen. The level of enhancement of bitumen with NR depends on how properly NR is dispersed in it. The rate and quality of dispersion is dependent on mixing speed and duration. An increase in mixing speed also reduces the time needed for complete dispersion [4]. Other studies were conducted on the epoxidation of natural rubber before being added to bitumen. The epoxidation of NR latex is a chemical process where NR latex is modified with proxy formic acid. Epoxidized natural rubber (ENR) has better chemical and heat resistance as compared to pure NR due to its ability to resist crystallization at high glass transition temperatures [40]. The epoxidation of NR latex improves the adhesion and toughness of the material [44]. The following is a review of some studies conducted on NR in bituminous mixes.

1. Al-Mansob *et al.* [45] studied the physical and rheological properties of ENR-modified bitumen. They used 80/100 penetration grade bitumen and ENR with an epoxidation of 53% passing through a 2.36 mm sieve. Bitumen was heated till 160°C in a high-shear mixer at 4000 rpm speed. ENR was then added to the heated bitumen when it became stable at 160°C and stirred at 4000 rpm speed for an hour.

   The sample was then poured into an aluminum foil tube and kept vertically in an oven at a temperature under 163°C for 48 h. After that, the sample was left to cool before being cut into pieces to undergo initial testing. The samples were aged with a pressure-aging vessel and a rolling thin film oven. They were first placed in a rolling thin film oven at 163°C for 85 min and then placed in the pressure-aging vessel at 100°C for 20 h. The authors found that ENR stiffened the base bitumen and that the ductility of the mixture increased after certain portions of ENR was added. They concluded that 6% ENR in bitumen was most favorable to modify bitumen.

2. Cheriet *et al.* [46] studied the effect of NR on the creep property of bituminous concrete. They used rubber powder from industrial waste to modify bitumen to be used as concrete for road surfaces. Bitumen of penetration grade 35/50 was used for this experiment. NR for the experiment was industrial waste from a rubber company.

   The modification of bitumen was done by mixing portions of bitumen with varying portions of rubber in a mechanical stirrer at 193°C for 2 h at a speed of 600 rpm. The modified bitumen was then mixed with coarse and fine aggregates under varying proportions to form concrete. They concluded that modified bitumen with 4% and 8% NR produced higher results when it came to resistance against permanent deformation in asphalt concrete.
3. Shaffie *et al.* [43] further investigated the effect of mix variables on the physical properties of bitumen modified with NR latex. Bitumen was modified with 2%, 4%, 6%, 8%, and 10% weight portions of NR to see how they affect the physical properties of bitumen. Granite aggregate obtained from a quarry, NR latex, and 80/100 grade bitumen obtained from SHELL were used for this experiment. Polymer modification of bitumen was done in a high-shear mechanical mixer. The bitumen was melted at 110°C and poured into a container. Then the bitumen was heated in an oven till it liquified at 150°C. After that, NR latex was then added to the heated bitumen and stirred with a mechanical stirrer. When the temperature got to 160°C, the stirring speed was increased to 500 rpm for 30 min, while maintaining the temperature. For this experiment, two mixing temperatures (140°C and 160°C), two mixing times (60 min

and 30 min), and four stirring speeds (500 rpm, 1000 rpm, 1500 rpm, and 2000 rpm) were used. The authors found that including NR made bitumen more susceptible to temperature changes and concluded that 8% NR addition provided the most effective sample.

4. Kishore and Gottala [47] used NR to modify bitumen. They incorporated NR into bitumen in weight proportions of 1–7%. For this experiment, 70/100 grade bitumen and polyisoprene in balloon form were used. Melt blending was adopted for this experiment. Around 500 mg of bitumen was measured and melted in a container at temperatures between 150°C and 180°C. Then, the polymer is slowly added and stirred for 30 min to form a homogeneous mixture. The mixture was allowed to cool before undergoing tests.

   They found that bitumen with 2–4% binder content was optimal for stability. They reported that rubberized bitumen improved skid resistance, temperature susceptibility, and deformation resistance of road. They concluded that a road with rubberized bitumen could have its service life increased by 100% than roads with just bituminous mix.

5. Ansari *et al.* [40] reviewed the use of NR as bitumen modifier. In the study, they noted that NR is used in either latex or coagulated form. The interaction between NR and bitumen increased the viscoelasticity of bitumen. They also found that the strength of rubber-modified bitumen largely depends on the type of NR used. They found that the optimum proportion of NR in most papers was 4–6% of bitumen weight.

They reported that adding NR to bitumen was shown to improve fatigue and deformation resistance for pavement. Also, relying on NR for modification instead of petroleum-based modifiers is good for the environment and economical as NR is cost-effective. They suggested that in-depth research be done on aging at a low temperature of performance of NR-modified bitumen.

In another study, adding NR to bitumen was observed to decrease the workability of asphalt mixtures due to an increase in viscosity and adhesion [40]. Because of this, high temperature is required during construction for using asphalt, which means more energy would be needed during asphalt production. In addition to this, there is a concern for greenhouse gas emissions from the asphalt-making process because of high energy usage [48]. Some admixtures have been studied to produce warm mix asphalt to save energy and reduce greenhouse gas emissions. Abdulrahman *et al.* [49] studied the properties of PMB with Evotherm to produce warm mix asphalt. Evotherm is a chemical additive added to asphalt; it reduces the temperature for mixing to the range of about 50–75°C [50]. They found that adding the additive reduced the viscosity of the binder, thereby permitting a lower mix and compaction of the binder. Evotherm also considerably increased the softening point of the binder. They recommended 0.5% of Evotherm as an admixture for warm mix asphalt. Bindu *et al.* [4] evaluated the performance of warm mix asphalt modified with NR and cashew nut shell liquid (CNSL) as an additive. They found that the optimum quantity of CNSL that reduced viscosity was 2%. It reduced the viscosity of the binder by 44% at 120°C. Ashiru Sani *et al.* [51] investigated the effects of ZycoTherm on NR latex–modified asphalt binder. The binder with ZycoTherm showed an improvement in elasticity and performance at low temperatures. The stiffness of the binder with ZycoTherm was relatively better than the one without the additive. They concluded that although the performance of the binder with NR alone was good, the addition of ZycoTherm showed an improvement.

In addition to the fact that NR as bitumen modifier has proven to be an effective binder from laboratory testing. The elastomeric nature of the binder improved thermal sensitivity, creep and fatigue resistance, and good adhesion adds to what makes it exceptional from ordinary asphalt mixes. Pavements conducted with these binders showed a decrease in rutting behavior after 6 months of construction [52]. Ahazar *et al.* [21] reported that field testing of pavements with NR-modified asphalt as a binder performed excellently when subjected to load as compared to the one without. However, tests that show the performance of NR-modified asphalt pavements in the field are limited.

13.6 Challenges

Bitumen modification with NR is a very expensive process, resulting in the practice being neglected [53]. The modification of NR with bitumen requires high temperature and high energy, which makes it costly [54]. Another challenge with modifying bitumen with NR is the segregation of polymer in bitumen. NR in dry form when dispersed ineffectively in bitumen forms a heterogenous blend [53]. Read and Whiteoak [55] reported that the difference in polarity, solubility, and molecular weight of bitumen and polymer makes the materials thermodynamically incompatible. This causes delamination of PMB, which negatively affects the material [45]. To combat these problems, solvents such as toluene and nano-alumina are added to the mix to aid the dispersion of NR in bitumen. Ismail *et al.* [56] investigated the effect of toluene in PMB and concluded that toluene improved the compatibility of the polymer and bitumen and did not adversely affect the performance of the product.

The farming of rubber trees in most areas continues to reduce in this modern age because most farmers have moved from rubber production to cash crops such as cocoa, cashew, palm oil, and others [57]. The demand for NR has been declining in recent times because the market is flooded with cheap synthetic rubber, which also has good qualities.

13.7 Conclusion and Recommendations

The aim of this study is to review the literature on the feasibility of NR in bituminous mixes for road and pavement construction. NR, as noted from the review, is an eco-friendly material, and promoting its consumption instead of petroleum-based alternatives is a step toward sustainable development. It has been proven that NR can be used to reduce the difficulty in the conventional asphalt-mixing process and produce a better performing binder. Thus, the influence of NR on the properties of bitumen mixes is significant to achieve high-quality mixes. For the road and pavement industry, the evidence points to the fact that NR can be used in construction to make sustainable pavements and roads. The increase in viscoelasticity of the binder helps improve the service life of pavements while reducing maintenance cost.

The effect of NR on bitumen is highly dependent on the type of NR used. The change in bitumen with NR is also dependent on factors such as the mixing speed and temperature during blending. The optimum value of NR that can be effectively incorporated into bitumen is between 4% and 8% of bitumen weight, as per the literature reviewed. While NR is an eco-friendly material, it absorbs an estimated 24.6 kg of CO_2 to produce 1 kg of NR latex, representing a material that supports the mitigating effect of climate change. However, there have not been enough tests on the performance of NR-modified bitumen in the field that give accurate results on how it affects the nature and service life of pavements and roads. The addition of NR to asphalt mixes causes a reduction in the workability of the product, which subsequently requires higher mixing and compaction temperatures. Although there are additives that can be used to get warm mix asphalt to reduce the mixing and compaction temperatures, this adds to the cost of production.

References

1. Nair, K.P.P. 2010. Rubber (*Hevea brasiliensis*), in *The Agronomy and Economy of Important Tree Crops of the Developing World*, Elsevier, pp. 237–273. https://doi.org/10.1016/B978-0-12-384677-8.00008-4.
2. Al-Sabaeei, A., Yussof, N.I.M., Napiah, M., and Sutanto, M. 2019. A review of using natural rubber in the modification of bitumen and asphalt mixtures used for road construction. *Jurnal Teknologi*, **81**(6).
3. Abdulrahman, S., Hainin, M.R., Satar, M.K.I.M., Hassan, N.A., and Al Saffar, Z.H. 2020. Review on the potentials of natural rubber in bitumen modification, in *IOP Conference Series: Earth and Environmental Science* (Vol. 476, No. 1), IOP Publishing, p. 012067.
4. Bindu, C.S., Joseph, M.S., Sibinesh, P.S., George, S., and Sivan, S. 2020. Performance evaluation of warm mix asphalt using natural rubber modified bitumen and cashew nut shell liquid. *International Journal of Pavement Research and Technology*, **13**(4), 442–453.
5. Appiah, J.A., Berko-Boateng, V.N., and Tagbor, T.A. 2016. Use of waste plastic materials for road construction in Ghana. *Case Studies in Construction Materials*, **6**, 1–7. https://doi.org/10.1016/j.cscm.2016.11.001.
6. Cornish, K. 2017. Alternative natural rubber crops: Why should we care? *Technology & Innovation*, **18**(4), 244–255.
7. Mooibroek, H., and Cornish, K. 2000. Alternative sources of natural rubber. *Applied Microbiology and Biotechnology*, **53**(4), 355–365.
8. Buranov, A., and Elmuradov, B.J. 2010. Extraction and characterization of latex and natural rubber from rubber-bearing plants. *Journal of Agricultural and Food Chemistry*, **58**(2), 734–743.

9. Venkatachalam, P., Geetha, N., Sangeetha, P., and Thulaseedharan, A. 2013. Natural rubber producing plants: An overview. *African Journal of Biotechnology*, **12**(12).

10. Yamashita, S., and Takahashi, S. 2020. Molecular mechanisms of natural rubber biosynthesis. *Annual Review of Biochemistry*, **89**, 821–851.

11. Oluwole, O.I., Avwerosuoghene, O.M., and Oluwatobi, A.J. 2015. The effect of natural rubber on the flexural properties of coconut coir (*Cocos nucifera*) reinforced red sand composites. *Acta Technica Corviniensis-Bull Eng*, **8**, 87–92.

12. Tanaka, Y. 2001. Structural characterization of natural polyisoprenes: Solve the mystery of natural rubber based on structural study. *Rubber Chemistry and Technology*, **74**(3), 355–375.

13. Intergovernmental Panel on Climate Change (IPCC). 2014. Summary for Policymakers. In: *Climate Change 2013 – The Physical Science Basis: Working Group I Contribution to the Fifth Assessment Report of the Intergovernmental Panel on Climate Change*, Cambridge University Press, pp. 1–30.

14. Hirata, Y., Kondo, H., and Ozawa, Y. 2014. Natural rubber (NR) for the tyre industry, in *Chemistry, Manufacture and Applications of Natural Rubber*, Woodhead Publishing, pp. 325–352.

15. Khan, I., and Poh, B.T. 2011. Natural rubber-based pressure-sensitive adhesives: A review. *Journal of Polymers and the Environment*, **19**(3), 793–811.

16. Shaffie, E., Ahmad, J., Arshad, A.K., Kamarun, D., and Kamaruddin, F. 2015. Stripping performance and volumetric properties evaluation of hot mix asphalt (HMA) mix design using natural rubber latex polymer modified binder (NRMB), in *InCIEC* 2014, Springer, Singapore, pp. 873–884.

17. Escobar, C.F., and Dos Santos, L.A. 2015. New eco-friendly binder based on natural rubber for ceramic injection molding process. *Journal of the European Ceramic Society*, **35**(13), 3567–3575.

18. Rakesh, K., Vijayan, S., Wilson, P., Kumar, P.A., and Prabhakaran, K. 2019. Aqueous tape casting of alumina using natural rubber latex binder. *Ceramics International*, **45**(15), 18543–18550.

19. Zhou, Y., Kosugi, K., Yamamoto, Y., and Kawahara, S. 2017. Effect of non-rubber components on the mechanical properties of natural rubber. *Polymers for Advanced Technologies*, **28**(2), 159–165.

20. Lake, G.J., Samsuri, A., Teo, S.C., and Vaja, J. 1991. Time-dependent fracture in vulcanized elastomers. *Polymer*, **32**(16), 2963–2975.

21. Azahar, N.F.B.M., Hassan, N.A., Jaya, R.P., Kadir, M.A.B.A., Yunus, N.Z.B.M., and Mahmud, M.Z.H. 2016. An overview on natural rubber application for asphalt modification. *International Journal of Agriculture, Forestry and Plantation*, **2**(Feb), 212–218.
22. Wiranata, A., and Malik, A. 2021. The effect of technical natural rubber mastication with wet process mixing on the characteristics of Asphalt-Rubber Blend, in *Journal of Physics: Conference Series* (Vol. 2049, No. 1). IOP Publishing, p. 012081.
23. Thuong, N.T., Yamamoto, O., Nghia, P.T., Cornish, K., and Kawahara, S. 2017. Effect of naturally occurring crosslinking junctions on green strength of natural rubber. *Polymers for Advanced Technologies*, **28**(3), 303–311.
24. Krishnapriya, M.G. 2015. Performance evaluation of natural rubber modified bituminous mixes. *International Journal of Civil, Structural, Environmental and Infrastructure Engineering Research and Development* (IJCSEIERD), **5**(34), 121–134.
25. Phinyocheep, P. 2014. Chemical modification of natural rubber (NR) for improved performance, in *Chemistry, Manufacture and Applications of Natural Rubber*, Woodhead Publishing, pp. 68–118.
26. Nawamawat, K., Sakdapipanich, J.T., Ho, C.C., Ma, Y., Song, J., and Vancso, J.G. 2011. Surface nanostructure of *Hevea brasiliensis* natural rubber latex particles. *Colloids and Surfaces A: Physicochemical and Engineering Aspects*, **390**(1–3), 157–166.
27. Manshaie, R., Khorasani, S.N., Veshare, S.J., and Abadchi, M.R. 2011. Effect of electron beam irradiation on the properties of natural rubber (NR)/styrene–butadiene rubber (SBR) blend. *Radiation Physics and Chemistry*, **80**(1), 100–106.
28. Matos, C.F., Galembeck, F., and Zarbin, A.J. 2014. Multifunctional and environmentally friendly nanocomposites between natural rubber and graphene or graphene oxide. *Carbon*, **78**, 469–479.
29. Lee, S.H., Lee, J.H., Nam, D.H., Cho, M., Kim, J., Chanthad, C., and Lee, Y. 2018. Epoxidized natural rubber/chitosan network binder for silicon anode in lithium-ion battery. *ACS Applied Materials & Interfaces*, **10**(19), 16449–16457.
30. Redelius, P., and Soenen, H. 2015. Relation between bitumen chemistry and performance. *Fuel*, **140**, 34–43.
31. Weigel, S., and Stephan, D. 2018. Relationships between the chemistry and the physical properties of bitumen. *Road Materials and Pavement Design*, **19**(7), 1636–1650.

32. Porto, M., Caputo, P., Loise, V., Eskandarsefat, S., Teltayev, B., and Oliviero Rossi, C. 2019. Bitumen and bitumen modification: A review on latest advances. *Applied Sciences*, **9**(4), 742.
33. Saad, A.A., Ismail, Y., Madi, H., Marwan, A.B., and Abdalla, S.A. 2013. Pavement performance with carbon black and natural rubber (latex). *International Journal of Engineering and Advanced Technology* (IJEAT), **2**(3), 124–129.
34. Tauste, R., Moreno-Navarro, F., Sol-Sánchez, M., and Rubio-Gámez, M.C. 2018. Understanding the bitumen ageing phenomenon: A review. *Construction and Building Materials*, **192**, 593–609.
35. Lesueur, D. 2009. The colloidal structure of bitumen: Consequences on the rheology and on the mechanisms of bitumen modification. *Advances in Colloid and Interface Science*, **145**(1–2), 42–82.
36. Omar, A.H., Mubaraki, M., Abdullah, N.H., Al-Sabaeei, A., Memon, A.M., Imran Khan, M., Sutanto, M.H., and Md Yusoff, N.I. 2021. Determining the effect of ageing of nano-clay modified bitumen using atomic force microscopy. *Scientia Iranica*, **28**(5), 2534–2545.
37. Arabani, M., and Esmaaeli, N. 2018. Laboratory evaluation on effect of groundnut shell ash on performance parameters of asphalt binder and mixes. *Road Materials and Pavement Design*, **21**(5), 1–23.
38. Al-Mansob, R.A., Ismail, A., Md Yusoff, N.I., Azhari, C.H., Karim, M.R., Alduri, A., and Baghini, M.S. 2014. Rheological characteristics of epoxidized natural rubber modified bitumen, in *Applied Mechanics and Materials* (Vol. 505), Trans Tech Publications Ltd, pp. 174–179.
39. Mustafa Kamal, M., Arifin Hadithon, K., and Abu Bakar, R. 2020. Natural rubber modified asphalt, in *IOP Conference Series: Earth and Environmental Science* (Vol. 498), IOP Publishing, p.012001. https://doi.org/10.1088/1755-1315/498/1/012001.
40. Ansari, A.H., Jakarni, F.M., Muniandy, R., Hassim, S., and Elahi, Z. 2021. Natural rubber as a renewable and sustainable bio-modifier for pavement applications: A review. *Journal of Cleaner Production*, **289**, 125727.
41. Wang, S., Cheng, D., and Xiao, F. 2017. Recent developments in the application of chemical approaches to rubberized asphalt. *Construction and Building Materials*, **131**, 101–113. https://doi.org/10.1016/j.conbuildmat.2016.11.077.
42. Wen, Y., Wang, Y., Zhao, K., and Sumalee, A. 2017. The use of natural rubber latex as a renewable and sustainable modifier of asphalt binder. *International Journal of Pavement Engineering*, **18**(6), 547–559.

43. Shaffie, E., Arshad, A.K., Alisibramulisi, A., Ahmad, J., Hashim, W., Abd Rahman, Z., and Jaya, R.P. 2018. Effect of mixing variables on physical properties of modified bitumen using natural rubber latex. *International Journal of Civil Engineering and Technology*, **9**, 1812–1821.

44. Tanjung, F.A., Hassan, A., and Hasan, M. 2015. Use of epoxidized natural rubber as a toughening agent in plastics. *Journal of Applied Polymer Science*, **132**(29).

45. Al-Mansob, R.A., Ismail, A., Alduri, A.N., Azhari, C.H., Karim, M.R., and Yusoff, N.I.M. 2014. Physical and rheological properties of epoxidized natural rubber modified bitumens. *Construction and Building Materials*, **63**, 242–248.

46. Cheriet, F., Soudani, K., and Haddadi, S. 2015. Influence of natural rubber on creep behavior of bituminous concrete. *Procedia-Social and Behavioral Sciences*, **195**, 2769–2776.

47. Kishore, K.S.N., and Gottala, A. 2015. A study on effect of addition of natural rubber on the properties of bitumen & bituminous mixes. *International Journal of Science Technology & Engineering*, **2**(1), 206–212.

48. Behnood, A. 2019. Application of rejuvenators to improve the rheological and mechanical properties of asphalt binders and mixtures: A review. *Journal of Cleaner Production*, **231**, 171–182. https://doi.org/10.1016/j.jclepro.2019.05.209.

49. Abdulrahman, S., Hainin, M.R., Satar, M.K.I.M., Hassan, N.A., and Usman, A. 2019. Rutting and moisture damage evaluation of warm mix asphalt incorporating POFA modified bitumen. *International Journal of Engineering and Advanced Technology*, **9**(1).

50. Kuang, Y. 2012. *Evaluation of Evotherm as a WMA Technology Compaction and Anti-Strip Additive*, Iowa State University, Ames, Iowa.

51. Ashiru Sani, M.R.M.H., Shariff, K.A., Jamshidi, A., Ibrahim, A.H., and Poovaneshvaran, S. 2019. Engineering and microscopic characteristics of natural rubber latex modified binders incorporating silane additive. *International Journal of Pavement Engineering*, **21**(1), 1–10.

52. Othman, Z., Hainin, M.R., Warid, M.N.M., Idham, M.K., and Kamarudin, S.N.N. 2018. Cup lump modified asphalt mixture along jalan Kuala Lumpur-Kuantan, daerah Temerloh, Pahang, in MATEC Web of Conferences (Vol. 250), EDP Sciences, p. 02007.

53. Abdulrahman, S., Hainin, M.R., Satar, M.K.I.M., Hassan, N.A., and Usman, A. 2021. Mechanical performance and global warming potential of unaged warm cup lump modified asphalt. *Journal of Cleaner Production*, **297**, 126653.
54. Wang, H., Liu, X., Apostolidis, P., and Scarpas, T. 2018. Review of warm mix rubberized asphalt concrete: Towards a sustainable paving technology. *Journal of Cleaner Production*, **177**, 302–314.
55. Read, J., and Whiteoak, D. 2015. *The Shell Bitumen Handbook*, 6th edition, Thomas Telford Ltd.
56. Ismail, A., Al-Mansob, R.A., Yusoff, N.I.B.M., and Karim, M.R. 2012. Effect of toluene as disperser on the bitumen modified with epoxidized natural rubber with aging simulation. *Australian Journal of Basic and Applied Sciences*, **6**(12), 97–101.
57. Ali, M.F., Akber, M.A., Smith, C., and Aziz, A.A. 2021. The dynamics of rubber production in Malaysia: Potential impacts, challenges and proposed interventions. *Forest Policy and Economics*, **127**, 102449.

Chapter 14

Agro-Industrial Supplementary Cementitious Materials for Realizing Sustainable Development Goals of the UN and Net-Zero CO_2 Emission

Ahsan Waqar,[a] Naraindas Bheel,[a] Mehmet Serkan Kirgiz,[b] Muhammad Basit Khan,[a] and Muhammad Shoaib Mansoor[c]

[a]*Department of Civil and Environmental Engineering, Universiti Teknologi Petronas, Bandar Seri Iskandar, Tronoh, Perak, 32610, Malaysia*

[b]*Academia of ACMCEN, Beaverton, OR 97075, USA*

[c]*The University of Sargodha, Sargodha, Punjab, Pakistan, University Road, Sargodha, Punjab 40100, Pakistan*

nakres42@yahoo.com

The collection of by-products from agricultural and industrial processes made significant impact on the built environment. Therefore, this chapter aims at presenting the features of some cementitious materials from agricultural and industrial processes, such as millet husk ash, wheat straw ash, corncob ash, palm fuel

Handbook of Innovative Adhesive Technology: Biodegradable and Cost-Effective Binder for Built Environment and Engineering Sciences
Edited by Mehmet Serkan Kirgiz

ISBN 978-981-5129-51-9 (Hardcover), 978-1-003-59892-3 (eBook)
www.jennystanford.com

oil ash, groundnut shell ash, ground granulated blast furnace slag, limestone powder, silica fume, sewage sludge ash, and quarry dust.

All of the aforementioned supplementary cementitious materials are presented with reference to their material composition, and a comparison has been made of the by-products-based binder with cement to elaborate on its capabilities to be used as cement-replacement material in concrete. Results showed that the use of the by-products is very useful in terms of binding capacity, workability, strength gain, generating new chemical compounds, unit mass, water retention, thermal conductivity, reducing the CO_2 emission, and realizing the cleaner built environment.

14.1 Introduction

The accumulation of waste materials from agriculture and industrial processes is creating significant negative impact on the environment. The world is already facing climate change problems, and further increase in solid waste deposition without proper disposal always results in disturbing the natural ecosystem [1]. The organic compounds present in the agricultural waste are a constant threat to the environment because they release harmful gases into the atmosphere that cause global warming, and further they can disturb the soil structure if not properly disposed. In comparison, industrial waste accumulation in landfills results in a significant release of greenhouse gases as well as extensive chemical reactions through the leaching process, which can also contaminate groundwater [2]. Most of the materials that are regarded as based have found their applications in the concrete production. It is because in the production of concrete, the most important material is the binder or cement, and it is also expensive as compared to other materials [3–5]. Any other waste material that can replace the content of cement to some extent to save cost as well as reduce environment impact is regarded as supplementary cementitious material. The market of supplementary cementitious materials is growing, and it is expected to grow in the future, as shown in Fig. 14.1.

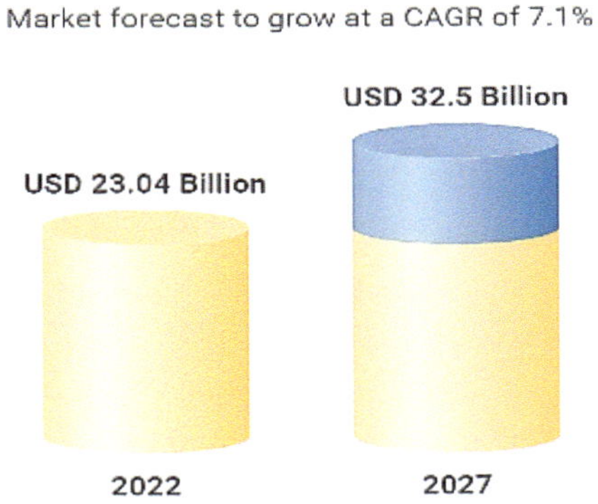

Figure 14.1 Expected growth in the supplementary cementitious materials market [6].

There is sustainability associated with the utilization of both agriculture and industrial waste in the production of concrete, due to which it also increases the economic advantage and helps reduce the negative impact on the environment. The chapter includes the presentation of millet husk ash, wheat straw ash, corncob ash, palm oil fuel ash, groundnut shell ash, ground granulated blast furnace slag (GGBFS), limestone powder, silica fume, sewage sludge ash, and quarry dust, as shown in Fig. 14.2. All of the above supplementary cementitious materials are presented with reference to their material composition and comparison with cement to elaborate their capabilities to be utilized as cement-replacement materials in concrete.

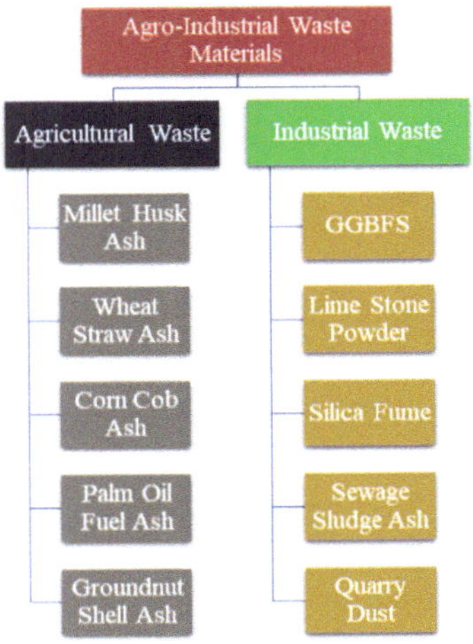

Figure 14.2 Agro-industrial waste materials.

14.2 Agricultural Wastes

14.2.1 Millet Husk Ash

Millet is widely grown in developing countries because of its significance as a major cereal-based food product. The annual production of millets is more than 28 million tons, out of which the African countries produce 40% because of their environment, which is supportive of the millet crop [7]. When the millet crop is harvested, the remaining millet husk goes to waste, utilized as an organic fertilizer, or burned. Millet husk itself does not have binding properties [8]. Still, after an uncontrolled burning process at 850°C, it produces millet ash, which is highly significant in terms of comparable properties with cement. Because of its presence in abundance, it can be utilized as an effective material to improve the compressive strength of concrete by more than 10% [9].

According to the existing research, increasing the quantity of millet husk in concrete always results in reducing the availability, but a sustainable percentage is found to be limited to 5% by weight of the concrete [10]. Table 14.1 shows the chemical composition of millet ash, according to which there is 74.08% silica content, while the binder calcium oxide is only 1.64%. It shows some level of burning properties, but because of high silica content, it reduces the burning properties compared to cement and provides greater resistance to handle compressive loads [11].

The material can be regarded as one of the suitable supplementary cementitious materials for developing countries where cement cost is high [12]. Millet husk effectively increments the compressive strength while also considering the cost factor in the construction of concrete, which can be reduced by 5% [13]. The uncontrolled burning process mostly used to produce millet husk ash is the major contributor to increasing the silica content while reducing the presence of calcium oxide, which is the always-required binder in standard cement, as shown in Fig. 14.3.

Table 14.1 Chemical composition of millet husk ash and standard Type I cement

Chemical Compounds	Millet Husk Ash Produced at 850°C (%)	ASTM Type I Cement (%)
Silicon dioxide (SiO_2)	74.08	20.4
Aluminum oxide (Al_2O_3)	1.66	5.2
Ferric oxide (Fe_2O_3)	4.17	3.3
Calcium oxide (CaO)	1.64	64.0
Magnesium oxide (MgO)	0.73	0.8
Sulfur trioxide (SO_3)	1.30	3.3

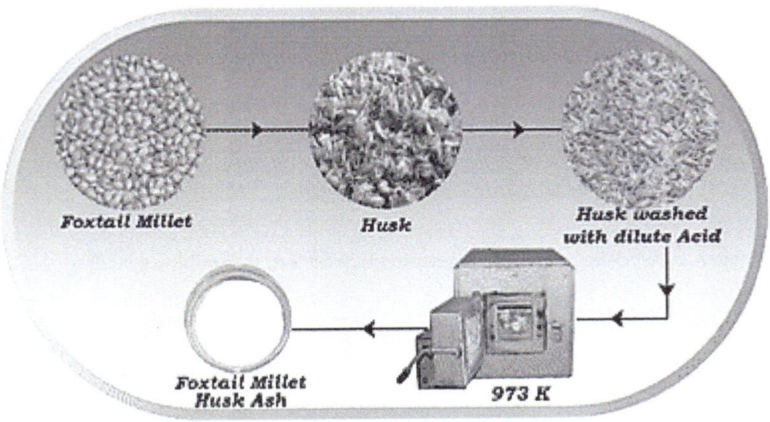

Figure 14.3 Production process of millet husk ash [8].

14.2.2 Wheat Straw Ash

After wheat grains are harvested, farmers usually burn the remaining stalk to produce wheat straw ash. Worldwide, cereal production is around 880 million tons, out of which more than 550 million tons of wheat straw are produced, as shown in Fig. 14.4 [14]. Wheat straw goes through total weight loss and thermal decomposition as organic waste material. Pozzolanic properties are present in wheat straw ash because of the presence of around 13% calcium oxide by weight of the ash.

Figure 14.4 Wheat straw ash produced at different temperatures [13].

The ash can be grounded to the required particle size for practical applications as pozzolanic material. It is suitable for supplementary cementitious material and satisfies the ASTM classes F, N, and C [15]. The temperature influences the pozzolanic properties of wheat straw ash. Normally for practical construction applications, it is produced in a temperature-controlled environment at 570°C, which gives the wheat straw ash a specific gravity of 2.31 [16].

If the temperature increases to 670°C, the specific gravity increases to 2.41 with variation in silica content. The chemical properties of wheat straw ash and standard cement are given in Table 14.2. The high silicon dioxide content promises high strength but also increases the setting time [17]. Compared with ordinary Portland cement, Table 14.2 shows that wheat straw ash produced at 670°C has a calcium oxide content of 12.54%, but in normal cement, it reaches 64%. Calcium oxide works as a binder when any pozzolanic material is used [18]. Its presence in high quantity makes the material show extensive bond strength.

Similarly, aluminum oxide helps the pozzolanic material achieve high compressive strength, and wheat straw ash and standard cement show minor differences in the quantity of aluminum oxide [19]. The chemical composition of wheat straw ash will not always remain the same, as shown in Table 14.2. Still, because of its ability to be used as supplementary cementitious material, it will always have significant application in the construction industry [20]. However, temperature, particle size, chemical composition, production technique, crystalline nanostructure, and the presence of organic compounds will always have a significant impact on the properties of wheat straw ash.

Table 14.2 Chemical composition of wheat straw ash and standard Type I cement

Chemical Compounds	Wheat Straw Ash Produced at 570°C (%)	Wheat Straw Ash Produced at 670°C (%)	ASTM Type I Cement (%)
Silicon dioxide (SiO_2)	73.06	83.8	20.4
Aluminum oxide (Al_2O_3)	3.90	4.55	5.2
Ferric oxide (Fe_2O_3)	1.75	1.05	3.3
Calcium oxide (CaO)	8.12	12.54	64.0
Magnesium oxide (MgO)	2.80	2.39	0.8
Sulfur trioxide (SO_3)	1.91	1.49	3.3

14.2.3 Corncob Ash

Corn is a crop widely grown in different countries with an appropriate warm climate, and corncob is the by-product after which a good part of the corn crop is harvested. It is estimated that the annual production of corncob is increasing, and it has reached the level of 164 T_g corncobs. More than 20% of corncob is produced in China only [21]. It is considered one of the significant agricultural wastes that always need proper solutions to help the environment get better. For utilization in the construction industry, corncob is burned above 500°C, producing the corncob ash with effective binding properties and high silica content [22].

It can be regarded as a medium supplementary cementitious material that shows binding properties when combined with cement [23]. Table 14.3 shows that the silica content is 65%, while up to 10% calcium oxide is present. It is an effective indication of binding properties because calcium oxide provides the critical strength comparable to cement when present in a high quantity, as shown in Fig. 14.5 [24]. Other chemicals, such as ferric oxide and aluminum oxide, are present in higher ranges than cement, which directly impacts the concrete-setting time if used. Corncob ash increases the compressive strength of concrete because of its high calcium oxide. It is also effective in using concrete where tensile forces are more dominant [25].

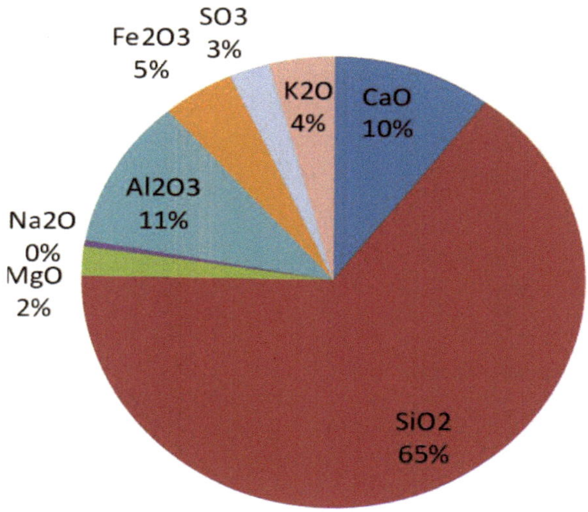

Figure 14.5 Typical material composition of corncob ash [8].

The variation in content can happen because of changes in the burning environment, which will always increase the silicon dioxide content in the material, which has negative implications for achieving high strength in concrete [26]. However, around 8% of the cost of concrete can be reduced if corncob ash is used. But its addition to concrete along with cement will always be limited to not more than 10% because it increases the silica content, which is bad for the overall strengthening of concrete over 28 days [27].

Table 14.3 Chemical composition of corncob and standard Type I cement

Chemical Compounds	Corncob Ash Produced at 500°C (%)	ASTM Type I Cement (%)
Silicon dioxide (SiO_2)	65.00	20.4
Aluminum oxide (Al_2O_3)	11.00	5.2
Ferric oxide (Fe_2O_3)	5.00	3.3
Calcium oxide (CaO)	10.00	64.0
Magnesium oxide (MgO)	2.00	0.8
Sulfur trioxide (SO_3)	3.00	3.3

14.2.4 Palm Fuel Oil Ash

The cultivation of palm crops has increased significantly over the past few decades, and now it provides more than 73 million metric tons of palm oil annually. As a major agricultural organic material, palm crops have significant applications as binding materials after being burned in a controlled environment or in palm oil production [28]. As a bio-product of the palm oil industry, it is always obtained by burning all sources of waste materials, including the fiber and different organic components of the harvested palm crop. It is estimated that around 110 million tons of waste is generated in the palm oil production industry, creating serious waste management issues [2]. Usually, the presence of silicon dioxide content is less than 60% in palm oil fuel ash after the burning process above 700°C, as shown in Fig. 14.6. The calcium oxide content cannot be increased by more than 10% in palm oil fuel because of limitations related to the process by which it is produced and the way it combines with the other chemicals produced in the burning process [29].

The presence of aluminum oxide and ferric oxide is in the acceptable range compared to cement. Table 14.4 shows that the silicon dioxide content is 53.82%, while the calcium oxide content is only 4.24% [30]. The material cannot be regarded as a high-grade supplementary cementitious material because it does not have the appropriate content of calcium oxide, which always shows expensive binding properties as in cement [31]. The chemical composition

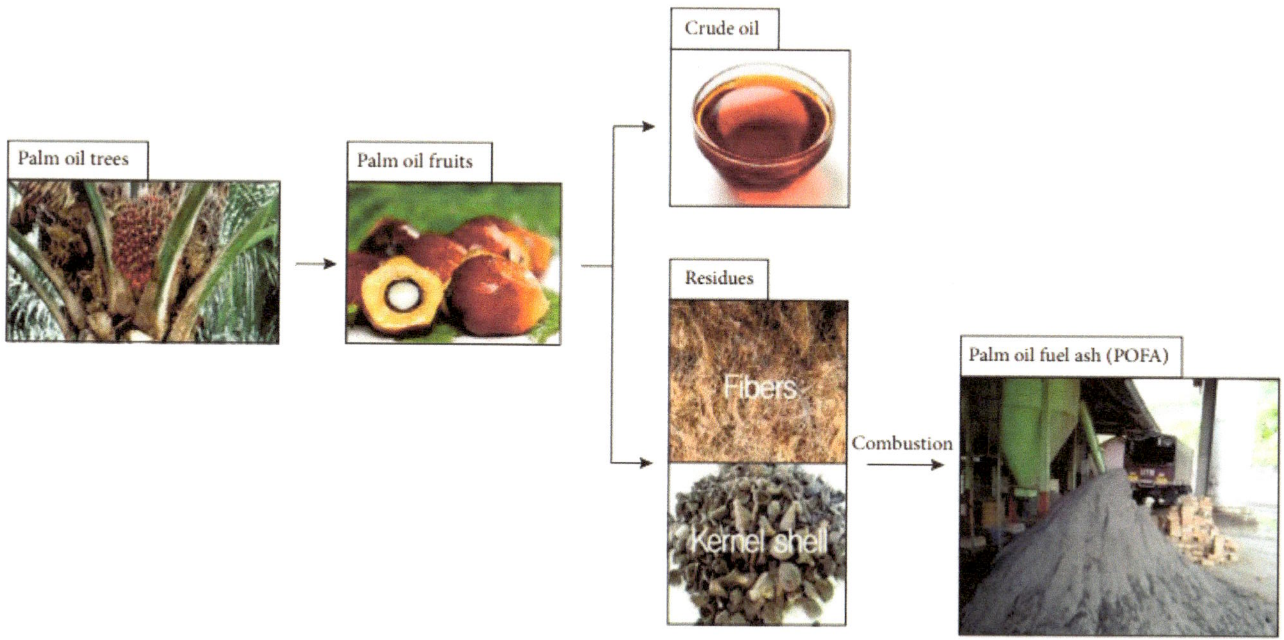

Figure 14.6 Production of palm oil fuel ash [8].

is highly variable because it depends on the material used in the production of palm oil fuel [32]. For example, if only fiber is used, it will have less impact on increasing the silicon dioxide content than the shells that are burned to produce the palm oil fuel ash [33]. However, it is important to reduce the cost of concrete while providing around 10% improvement in compressing the 10th and less than 5% improvement in the tensile resistance.

Table 14.4 Chemical composition of palm fuel oil ash and standard Type I cement

Chemical Compounds	Palm Oil Fuel Ash Produced at 700°C (%)	ASTM Type I Cement (%)
Silicon dioxide (SiO_2)	53.82	20.4
Aluminum oxide (Al_2O_3)	5.66	5.2
Ferric oxide (Fe_2O_3)	4.54	3.3
Calcium oxide (CaO)	4.24	64.0
Magnesium oxide (MgO)	3.19	0.8
Sulfur trioxide (SO_3)	2.25	3.3

14.2.5 Groundnut Shell Ash

Groundnut shells always contain organic functional compounds, which are highly significant for providing a better chemical composition comparable to cement. The annual production of groundnut in the world is more than 71 million metric tons, and around 23% of the net production goes to waste in the form of groundnut shells [1]. Some of it is utilized in the agriculture industry as an organic fertilizer, but most of it remains stable in the soil.

The controlled burning of groundnut shells gives groundnut shell ash, and the temperature is always maintained at 600°C [34]. Groundnut shell ash is always acceptable as one of the significant materials with a high content of calcium oxide because of organic functional groups, making it a good binder. Silicon dioxide content is also less in the ash produced from groundnut shells because of the natural deficiency of silicon in groundnut shells as they are harvested, as shown in Fig. 14.7 [35].

Figure 14.7 Groundnut shells and produced ash by burning [8].

Increasing the temperature further while making the groundnut shell for construction purposes will not increase the calcium oxide content because of the limited elements present in the groundnut shell's organic functional groups [36]. From Table 14.5, it is evident that the silicon dioxide content is 45.42%, less than 50%, making it a good material. This is because the high presence of silicon dioxide can directly interfere with the binding properties of the material, and it will eventually result in degrading the quality of concrete after the material is added along with cement [37].

Table 14.5 Chemical composition of groundnut shell ash and standard Type I cement

Chemical Compounds	Groundnut Shell Ash Produced at 600°C (%)	ASTM Type I Cement (%)
Silicon dioxide (SiO_2)	41.20	20.4
Aluminum oxide (Al_2O_3)	11.75	5.2
Ferric oxide (Fe_2O_3)	12.60	3.3
Calcium oxide (CaO)	11.23	64.0
Magnesium oxide (MgO)	3.51	0.8
Sulfur trioxide (SO_3)	0.44	3.3

Calcium oxide is the most powerful chemical, which is present at up to 11.23%, making it a very significant binder, but its binding strength is nearly three times less than that of normal cement [38]. Its utilization in the production of concrete is evident up to the maximum of 10%, after which negative effects on the workability and also resist the issues of law concrete strength because of the presence of silicon dioxide [39]. The cost of cement can also be saved by up to 10% if the groundnut shell ash is added to the concrete. The material effectively qualifies as a supplementary cementitious material having significant applications in the construction industry, where it is present in abundance as a waste product.

14.3 Industrial Wastes

14.3.1 Ground Granulated Blast Furnace Slag

Blast furnaces are commonly used in industrial smelting processes to produce metals. After completing every function in the blast furnace, the best product remains at the bottom, known as the GGBFS, as shown in Fig. 14.8 [40]. The material is highly composed of metal compounds because of its nature of production. It is mostly observed that the silicate and aluminium are always in higher proportion compared to other compounds. The annual production of the material ranges from more than 25 million tons annually, and more than 50% is now being used in the construction industry for material development-related purposes [41]. GGBFS is known to make high compressive strength improvements in concrete when added as a replacement material for cement because of the increased content of calcium oxide [42].

As an industrial waste, proper disposal is always a major problem for GGBFS. Its applications are continuously increasing in the construction industry, which is also used for cement production. The silicon dioxide content is normally between 27% and 38%, while the calcium oxide content can be 34–43% [43]. Compared to the other rigid materials produced, it is created from the perspective of binding strength, as calcium oxide is present in large quantities [44]. Table 14.6 shows the material composition of GGBFS in which the silica content is 38% and the calcium oxide content is 43%.

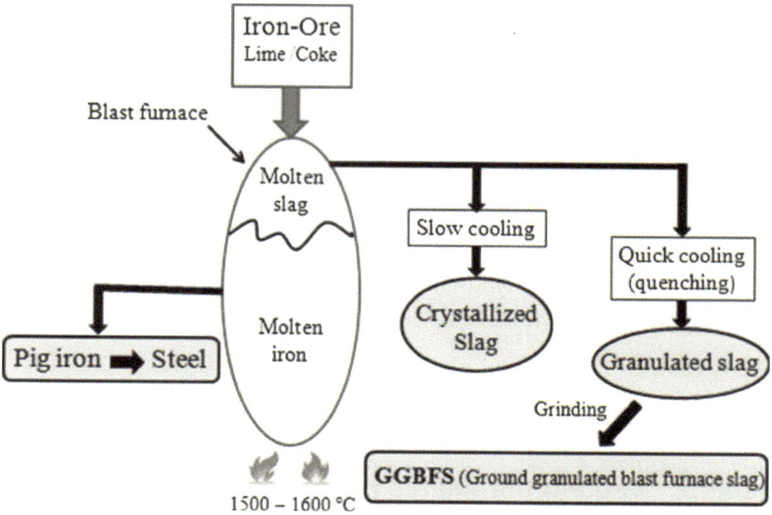

Figure 14.8 Production of GGBFS from iron smelting [23].

Further, the content of alumina is 7%, while the ferric oxide content reaches 1.6%. It has less impact on the setting time, due to which its ultimate mechanical properties are linked with improving compressive strength [45]. Because of its associated sustainability to the conception industry, it can be regarded as one of the highly significant supplementary cementitious materials, with a high content of calcium oxide binder that is also present in the same range in cement [46]. It can be replaced in concrete by more than 15%, according to requirements. Still, if increased, it also harms the workability and overall durability of concrete compared to cement.

Table 14.6 Chemical composition of GGBFS and standard Type I cement

Chemical Compounds	GGBFS (%)	ASTM Type I Cement (%)
Silicon dioxide (SiO_2)	38.00	20.4
Aluminum oxide (Al_2O_3)	7.00	5.2
Ferric oxide (Fe_2O_3)	1.60	3.3
Calcium oxide (CaO)	43.00	64.0
Magnesium oxide (MgO)	0.76	0.8
Sulfur trioxide (SO_3)	0.07	3.3

14.3.2 Limestone Powder

Limestone is mined from natural quarries, and its annual production is more than 8.8 billion metric tons globally. The production of limestone powder is a complete industrial process in which it is produced with the help of grinding equipment in which limestone is ground to fine powder [47]. The chemical composition of limestone consists of a very high percentage of calcium oxide, which is always expected to behave similarly to cement, as shown in the SEM analysis in Fig. 14.9. Its properties are comparable to cement because the rate of calcium oxide may reach 90% sometimes [48]. This makes it a pure binder, but when it is alone without cement, it cannot provide the required compressive strength because of the low percentage of silicon dioxide and other important compounds such as alumina and ferric oxide [49]. In most countries, it is allowed to add limestone to prepare cement clinker, and it also provides valuable efficiency to concrete when added up to 5% [50].

Table 14.7 shows the material composition of GGBFS in which the silica content is 38% and the calcium oxide content is 43%. The compressive strength slightly increases, but it offers greater resistance to bleeding and crack formation. Its placement in the cement is not increased by more than 5% because of the high content of calcium oxide that affects the chemical composition of cement and its compounds in concrete [51]. According to the following table, it is evident that the percentage of calcium oxide is 71.39, which is the highest compared to any supplementary cementitious material previously presented in this chapter. This is because limestone has a high percentage of calcium oxide, which is the same as grounded fine powder [52]. The most interesting thing here is the very low content of silicon dioxide, which is 8.2% and not enough to provide a valuable increment in compressive strength as its percentage is increased as a replacement material in concrete [53].

Similarly, the content of ferric oxide and alumina is also very low, which does not contribute to setting up appropriate timing for concrete. Also, it directly impacts reduced efficiency when the replacement percentage is increased above 5%. However, it can greatly reduce the cost of concrete production by more than 10% because of its abundant availability in the market.

Figure 14.9 Comparison between limestone powder (a) and cement (b) in SEM analysis [40].

Table 14.7 Chemical composition of limestone powder and standard Type I cement

Chemical Compounds	Limestone Powder (%)	ASTM Type I Cement (%)
Silicon dioxide (SiO_2)	8.20	20.4
Aluminum oxide (Al_2O_3)	1.53	5.2
Ferric oxide (Fe_2O_3)	0.96	3.3
Calcium oxide (CaO)	71.39	64.0
Magnesium oxide (MgO)	1.27	0.8
Sulfur trioxide (SO_3)	0.71	3.3

14.3.3 Silica Fume

Silica fume production is carried out in electrical arc furnaces, where the development of silicon-based material is carried out. Quartz crystals are heated above 200°C, which results in most silicon that ends up producing silica fume as a by-product, as shown in Fig. 14.10 [54]. The global production of silica is above 7 million metric tons, and almost all of this is utilized in different applications [55]. Its use in concrete started more than two decades ago because of its fine characteristics and appropriate particles, which improve the structure of concrete [56]. Silica fume is a highly acceptable material from the perspective of supplementary cementitious characteristics because it contains some degree of calcium oxide but high silicon dioxide content, which provides an appropriate improvement in concrete's compressive strength when added below 10% by weight of the cement. It has also been observed that silica fume can behave similarly to cement when combined with sodium hydroxide in concrete [57].

Silica fume does not have high binding characteristics. Still, it can be created as a low-grade binding material because less calcium oxide and more than 90% silicon dioxide are present. But when combined with cement, its characteristics are enhanced because the calcium oxide in cement reacts with the compounds present in silica [58]. It ultimately results in improving compressive strength and other durability characteristics of concrete. Table 14.8 shows that the silicon dioxide content is 92.03%, which is very high [59].

The calcium oxide content is less than 1%, meaning that the binding characteristics are almost negligible. Because of its presence as a by-product of industrial processes, it is widely utilized in combination with normal cement to take advantage of the high silicon dioxide content and improve compressive strength by more than 15% [60].

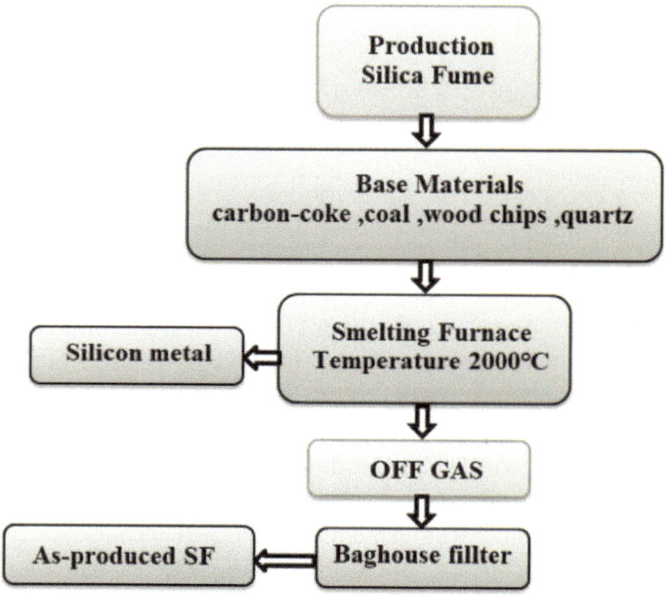

Figure 14.10 Industrial production in silica fume [8].

Table 14.8 Chemical composition of silica fume and standard Type I cement

Chemical Compounds	Silica Fume (%)	ASTM Type I Cement (%)
Silicon dioxide (SiO_2)	92.03	20.4
Aluminum oxide (Al_2O_3)	1.10	5.2
Ferric oxide (Fe_2O_3)	1.60	3.3
Calcium oxide (CaO)	0.70	64.0
Magnesium oxide (MgO)	2.10	0.8
Sulfur trioxide (SO_3)	0.85	3.3

14.3.4 Sewage Sludge Ash

Water treatment always results in the production of sewage sludge, which is bad for the environment and needs proper disposal or further treatment. As an industrial by-product, it is further processed by burning to form sewage sludge ash [61]. It is a fine material because of the combustion process in the preparation, and the temperature is always kept at about 800°C [62]. The utilisation of waste sewage sludge in concrete is witnessing a notable surge due to its environmental system efficacy and the presence of assorted chemical compounds resulting from the coagulants employed in wastewater treatment facilities, as illustrated in Fig. 14.11 [63]. There is always a high content of silicon dioxide and other important elements that provide unique characteristics to the waste material. It is also known that the sludge-conditioning process significantly impacts the ultimate chemical composition of the material, which has a variable impact on concrete's compressive strength [64]. However, the presence of silicon dioxide content will always be greater as compared to all other chemical compounds as it is the ultimate end product of the combustion process.

It has been observed that while using concrete, the material results in increasing the compressive strength but not above 8%. It is also indicated that concrete's effective replacement ratio from the perspective of cement lies between 5% and 10% [65]. Table 14.9 shows that the silicon dioxide content in the waste sewage sludge ash is 39.52%, and the next important material compound is aluminum oxide, which is 17.17%. The high presence of alumina greatly affects the sustainability of concrete production because it creates a prolonged setting time, which may not be suitable for the concrete [66]. This is a reason that a higher percentage of replacement of this material is not effective in concrete. But an appropriate increment in the compressive strength and a reduction in the cost of concrete can happen. Normally, the material reduces the cost of cement by more than 8% [67].

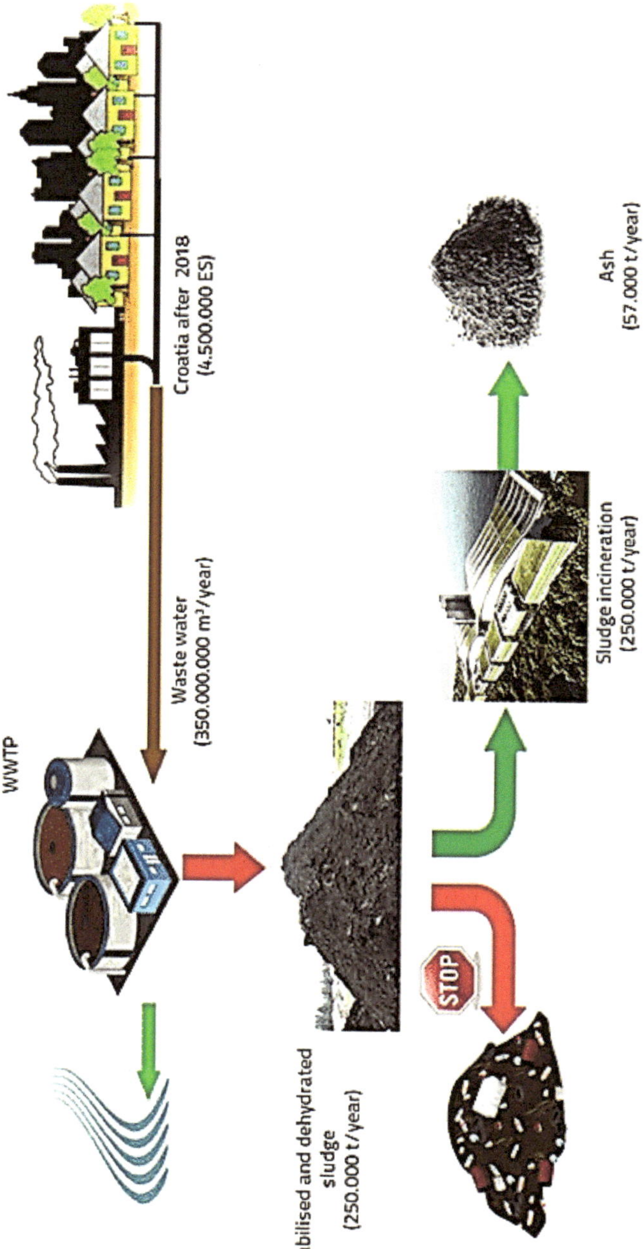

Figure 14.11 Sewage sludge ash production process [8].

Table 14.9 Chemical composition of sewage sludge ash and standard Type I cement

Chemical Compounds	Sewage Sludge Ash (%)	ASTM Type I Cement (%)
Silicon dioxide (SiO_2)	39.52	20.4
Aluminum oxide (Al_2O_3)	17.17	5.2
Ferric oxide (Fe_2O_3)	11.99	3.3
Calcium oxide (CaO)	7.16	64.0
Magnesium oxide (MgO)	2.13	0.8
Sulfur trioxide (SO_3)	1.8	3.3

14.3.5 Quarry Dust

Quarrying operations are carried out worldwide to mine stone aggregates, producing quarry dust, considered a waste material. In the process of quarrying, dust is also produced, known as one of the important supplementary cementitious materials. In the presence of moisture, the quarry dust may not provide relevant results when added to concrete as a replacement material [68]. The normal range of humidity is usually less than 10%, and the material is utilized in various concreting applications because of its relativity with the stone aggregate [69]. The average percentage of calcium oxide in quarry dust is usually less than 10%. The rocks utilized for aggregate production do not always have a high calcium oxide content. Further, it is not passed through the burning process, due to which the original material composition and grading remain the same as in the rock, as shown in Fig. 14.12 [3]. Table 14.10 shows the chemical composition of quarry dust and standard Type I cement.

Additionally, it is observed that the silicon dioxide content is above 60%, which gives a characteristic improvement in the compressive strength when combined with cement, but alone it cannot provide greater compressive strength because of a lower percentage of binding material [4]. Quarry dust is known to be highly sustainable because of its relativity with the stone aggregate, and it is also present in abundance, which does not affect the supply of construction materials [5].

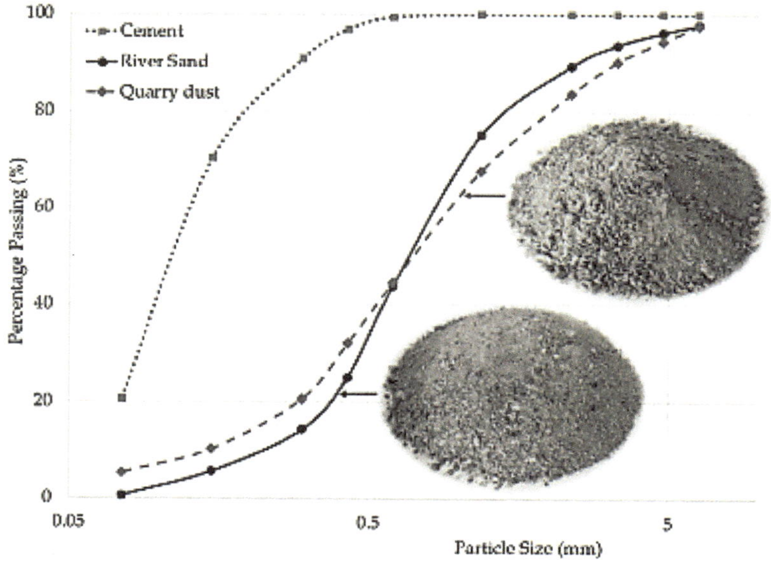

Figure 14.12 Comparison between the grading of quarry dust and cement [53].

According to the material composition given in Table 14.10, it is evident that the silicon dioxide content is 65.73%, while the alumina content is 19.31%. The high percentage of alumina content directly affects the concrete-setting time, and it has direct implications on durability [6]. The calcium oxide content is only 3.64%, so its binding capabilities can be graded low. However, its utilization is possible in concrete because of its ability to save the cost associated with cement consumption [70]. More than 7% is not suitable for concrete when the quarry dust is used as a cement-replacement material. This is because the higher percentage impacts the setting time and further decreases workability.

Table 14.10 Chemical composition of quarry dust and standard Type I cement

Chemical Compounds	Quarry Dust (%)	ASTM Type I Cement (%)
Silicon dioxide (SiO_2)	65.73	20.4
Aluminum oxide (Al_2O_3)	19.31	5.2
Ferric oxide (Fe_2O_3)	5.27	3.3
Calcium oxide (CaO)	3.64	64.0
Magnesium oxide (MgO)	2.16	0.8
Sulfur trioxide (SO_3)	0.11	3.3

References

1. N. Bheel, P. Awoyera, T. Tafsirojjaman, N. Hamah Sor, and S. Sohu. Synergic effect of metakaolin and groundnut shell ash on the behavior of fly ash-based self-compacting geopolymer concrete. *Constr. Build. Mater.*, 2021, **311**(1), 125327.
2. E. M. Opiso, C. B. Tabelin, C. V. Maestre, J. P. J. Aseniero, I. Park, and M. Villacorte-Tabelin. Synthesis and characterization of coal fly ash and palm oil fuel ash modified artisanal and small-scale gold mine (ASGM) tailings-based geopolymer using sugar mill lime sludge as Ca-based activator. *Heliyon*, 2021, **7**, e06654.
3. K. Sundaralingam, A. Peiris, A. Anburuvel, and N. Sathiparan. Quarry dust as river sand replacement in cement masonry blocks: Effect on mechanical and durability characteristics. *Materialia*, 2022, **21**, 101324.
4. M. M. Aliyu, M. M. Nuruddeen, and Y. A. Nura. The use of quarry dust for partial replacement of cement in cement-sand mortar. *Fudma J. Sci.*, 2021, **4**(4), 432–437.
5. K. C. Onyelowe, F. Sosa, M. E. Onyia, A. Eberemu, D. B. Van, F. D. A. Onyelowe, C. Igboayaka, F. Orji, M. O. Idrees, A. B. Salahudeen, T. Amhadi, K. Ibe, M. N. Van, C. Ikpa, K. Osinubi, B. Ugorji, L. Dao-Phuc, and M. N. Xuan. An experimental study on compaction behavior of lateritic soils treated with quarry dust based geopolymer cement. *J. Solid Waste Technol. Manag.*, 2021, **47**(1), 104–119.
6. Y. Zhao, J. Qiu, J. Xing, and X. Sun. Recycling of quarry dust for supplementary cementitious materials in low carbon cement. *Constr. Build. Mater.*, 2020, **237**, DOI: 10.1016/j.conbuildmat.2019.117608.
7. N. Bheel, P. Awoyera, I. A. Shar, S. A. Abbasi, S. H. Khahro, and A. K. Prakash. Synergic effect of millet husk ash and wheat straw ash on the fresh and hardened properties of Metakaolin-based self-compacting geopolymer concrete. *Case Stud. Constr. Mater.*, 2021, **15**, e00729.
8. O. O. Ugwu, T. T. Tarzomon, and G. Atom. Cement replacement for sustainable concrete: An overview. *Int. J. Innov. Res. Dev.*, 2018, **7**(5).
9. N. Bheel, M. O. A. Ali, Tafsirojjaman, S. H. Khahro, and M. A. Keerio. Experimental study on fresh, mechanical properties and embodied carbon of concrete blended with sugarcane bagasse ash, metakaolin, and millet husk ash as ternary cementitious material. *Environ. Sci. Pollut. Res.*, 2022, **29**(4), 5224–5239.
10. J. Wasiu and F. A. Oluwatosin. Experimental investigation of the pozzolanic potentials of millet and rice husk ash as mineral additive in self-compacting concrete. *Acta Polytech.*, 2020, **60**(4), 359–368.

11. N. Bheel, F. A. Memon, and S. L. Meghwar. Study of fresh and hardened properties of concrete using cement with modified blend of millet husk ash as secondary cementitious material. *Silicon*, 2021, **13**, 4641–4652.
12. P. T. Ravichandran and M. Ranjith. Experimental study on soil stabilization using millet husk ash. *J. Adv. Res. Dyn. Control Syst.*, 2018.
13. N. Bheel, F. A. Memon, and S. L. Meghwar. Study of fresh and hardened properties of concrete using cement with modified blend of millet husk ash as secondary cementitious material. *Silicon*, 2021, **13**(12), 4641–4652.
14. S. Memon, I. Wahid, M. Khan, M. Tanoli, and M. Bimaganbetova. Environmentally friendly utilization of wheat straw ash in cement-based composites. *Sustainability*, 2018, **10**(5), 1322.
15. S. A. Memon, I. Wahid, M. K. Khan, M. A. Tanoli, and M. Bimaganbetova. Environmentally friendly utilization of wheat straw ash in cement-based composites. *Sustain.*, 2018, **10**(5), 1322.
16. A. Qudoos, H. G. Kim, Atta-ur-Rehman, I. K. Jeon, and J.-S. Ryou. Influence of the particle size of wheat straw ash on the microstructure of the interfacial transition zone. *Powder Technol.*, 2019, **352**, 453–461.
17. M. Amin, T. Murtaza, K. Shahzada, K. Khan, and M. Adil. Pozzolanic potential and mechanical performance of wheat straw ash incorporated sustainable concrete. *Sustainability*, 2019, **11**(2), 519.
18. S. A. Memon, U. Javed, M. Haris, R. A. Khushnood, and J. Kim. Incorporation of wheat straw ash as partial sand replacement for production of eco-friendly concrete. *Materials (Basel).*, 2021, **14**(8), 2078.
19. J. Ahmad, R. F. Tufail, F. Aslam, A. Mosavi, R. Alyousef, M. F. Javed, O. Zaid, and M. S. K. Niazi. A step towards sustainable self-compacting concrete by using partial substitution of wheat straw ash and bentonite clay instead of cement. *Sustainability*, 2021, **13**(2), 824.
20. N. Bheel, M. H. W. Ibrahim, A. Adesina, C. Kennedy, and I. A. Shar. Mechanical performance of concrete incorporating wheat straw ash as partial replacement of cement. *J. Build. Pathol. Rehabil.*, 2021, **6**(1), 4.
21. Saloni, Parveen, T. Pham, Y. Lim, S. Pradhan, Jatin, and J. Kumar. Performance of rice husk ash-based sustainable geopolymer concrete with ultra-fine slag and corn cob ash. *Constr. Build. Mater.*, 2021, 279, 122526.
22. M. Shakouri, C. L. Exstrom, S. Ramanathan, and P. Suraneni. Hydration, strength, and durability of cementitious materials incorporating untreated corn cob ash. *Constr. Build. Mater.*, 2020.

23. S. Oyebisi, A. Ede, F. Olutoge, and S. Ogbiye. Evaluation of reactivity indexes and durability properties of slag-based geopolymer concrete incorporating corn cob ash. *Constr. Build. Mater.*, 2020, **258**, 119604.

24. M. Tumba, O. Ofuyatan, O. Uwadiale, J. Oluwafemi, and S. Oyebisi. Effect of sulphate and acid on self-compacting concrete containing corn cob ash. *IOP Conf. Ser. Mater. Sci. Eng.*, 2018, **413**, 012040.

25. N. Bheel and A. Adesina. Influence of binary blend of corn cob ash and glass powder as partial replacement of cement in concrete. *Silicon*, 2021, **13**(5), 1647–1654.

26. C. M. Gradinaru, M. Barbuta, D. Babor, and A. A. Serbanoiu. Corn cob ash as sustainable puzzolanic material for an ecological concrete. *Bull. Transilv. Univ. Brasov*, 2018.

27. John. Utilization of corn cobs ash as cementitious and binary cementitious materials in concrete and cement-based composites: a review. *DRC Sustain. Futur. J. Environ. Agric. Energy*, 2021, 26–42.

28. H. M. Hamada, A. A. Al-attar, F. M. Yahaya, K. Muthusamy, B. A. Tayeh, and A. M. Humada. Effect of high-volume ultrafine palm oil fuel ash on the engineering and transport properties of concrete. *Case Stud. Constr. Mater.*, 2020.

29. M. Amran, G. Murali, R. Fediuk, N. Vatin, Y. Vasilev, and H. Abdelgader. Palm oil fuel ash-based eco-efficient concrete: A critical review of the short-term properties. *Materials.*, 2021.

30. H. M. Hamada, G. A. Jokhio, F. M. Yahaya, A. M. Humada, and Y. Gul. The present state of the use of palm oil fuel ash (POFA) in concrete. *Constr. Build. Mater.*, 2018, **175**, 26–40.

31. K. Adebayo Mujedu, M. A. Ab-Kadir, and M. Ismail. A review on self-compacting concrete incorporating palm oil fuel ash as a cement replacement. *Constr. Build. Mater.*, 2020, DOI: 10.1016/j.conbuildmat.2020.119541.

32. H. M. Hamada, B. S. Thomas, F. M. Yahaya, K. Muthusamy, J. Yang, J. A. Abdalla, and R. A. Hawileh. Sustainable use of palm oil fuel ash as a supplementary cementitious material: A comprehensive review. *J. Build. Eng.*, 2021, **40**.

33. W. Chalee, T. Cheewaket, and C. Jaturapitakkul. Enhanced durability of concrete with palm oil fuel ash in a marine environment. *J. Mater. Res. Technol.*, 2021, **13**, 128–137.

34. K. K. Alaneme, M. O. Bodunrin, and A. A. Awe. Microstructure, mechanical and fracture properties of groundnut shell ash and silicon carbide dispersion strengthened aluminium matrix composites. *J. King Saud Univ. Eng. Sci.*, 2018, **30**(1), 96–103.

35. L. Venkatesh, T. V. Arjunan, and K. Ravikumar. Microstructural characteristics and mechanical behaviour of aluminium hybrid composites reinforced with groundnut shell ash and B4C. *J. Brazilian Soc. Mech. Sci. Eng.*, 2019, **41**(7).

36. A. W. Abro, A. Kumar, M. A. Keerio, Z. H. Shaikh, N. Bheel, and A. A. Dayo. An investigation on compressive strength of concrete blended with groundnut shell ash. *Neutron*, 2021.

37. N. Bheel, O. G. Aluko, and A. R. Khoso. Synergistic and sustainable utilization of coconut shell ash and groundnut shell ash in ternary blended concrete. *Environ. Sci. Pollut. Res.*, 2022, **29**(18), 27399–27410.

38. E. V. Khandelwal. Strength parameters of concrete using groundnut shell ash and recycled concrete aggregate as partial substitution of Opc-43 and coarse aggregate. *Int. J. Civ. Eng. Technol.*, 2019.

39. M. Arabani and N. Esmaaeli. Laboratory evaluation on effect of groundnut shell ash on performance parameters of asphalt binder and mixes. *Road Mater. Pavement Des.*, 2020.

40. R. Hidayawanti, Y. Latief, Y. Rahayu, F. I. Wachid, and R. P. A. Widodo. Utilization of GGBFS as cement substitute to reduce production cost of construction project. *Int. J. GEOMATE*, 2021, **21**(86), 9–16.

41. C. M. Yun, M. R. Rahman, C. Y. W. Phing, A. W. M. Chie, and M. K. Bin Bakri. The curing times effect on the strength of ground granulated blast furnace slag (GGBFS) mortar. *Constr. Build. Mater.*, 2020.

42. Y. Zhao, J. Gao, Z. Xu, S. Li, X. Luo, and G. Chen. Long-term hydration and microstructure evolution of blended cement containing ground granulated blast furnace slag and waste clay brick. *Cem. Concr. Compos.*, 2021, **118**.

43. L. Wang and F. Aslani. Electrical resistivity and piezoresistivity of cement mortar containing ground granulated blast furnace slag. *Constr. Build. Mater.*, 2020, **263**.

44. S. Bui Truong, N. Nguyen Thi, and D. Nguyen Thanh. An experimental study on unconfined compressive strength of soft soil-cement mixtures with or without GGBFS in the Coastal Area of Vietnam. *Adv. Civ. Eng.*, 2020.

45. S. Afroz, Y. Zhang, Q. Dieu Nguyen, T. Kim, and A. Castel. Effect of limestone in general purpose cement on autogenous shrinkage of high strength GGBFS concrete and pastes. *Constr. Build. Mater.*, 2022, **327**, 126949–126949.

46. A. A. Shubbar, H. Jafer, M. Abdulredha, Z. S. Al-Khafaji, M. S. Nasr, Z. Al Masoodi, and M. Sadique. Properties of cement mortar incorporated high volume fraction of GGBFS and CKD from 1 day to 550 days. *J. Build. Eng.*, 2020.

47. E. T. Dawood and M. N. A. Jaraullah. The effect of lime stone powder as a partial replacement of fine aggregate on the properties of mortar produced from SCMs, in *AIP Conference Proceedings*, 2022.

48. G. H. Hamedi, M. Sohrabi, F. Sakanlou, and S. A. Tahami. Investigating the effect of various fillers on cohesive failure mechanism in asphalt mixtures. *Period. Polytech. Civ. Eng.*, 2020.

49. I. Jaambavi, J. Esther Baby, T. Akshaya, and G. Rajeswari. Mechanical properties of concrete by replacement of fly ash, GGBFS, lime stone powder. *Mater. Today Proc.*, 2021.

50. K. Siamardi. Optimization of fresh and hardened properties of structural light weight self-compacting concrete mix design using response surface methodology. *Constr. Build. Mater.*, 2022.

51. T. Ino, T. Saito, K. Suzuki, and T. Saeki. Hydration reaction and considerations for soils treated by blast-furnace slag, anhydrite and lime stone powder. *Cem. Sci. Concr. Technol.*, 2021.

52. E. Sakai, M. Umetsu, N. Nito, and R. Kuga. Hydration of fly ash cement containing high alite cement and lime stone powder. *Cem. Sci. Concr. Technol.*, 2020.

53. N. Subash, S. Avudaiappan, S. Adish Kumar, M. Amran, N. Vatin, R. Fediuk, and R. Aepuru. Experimental investigation on geopolymer concrete with various sustainable mineral ashes. *Materials (Basel).*, 2021.

54. A. O. Habib, I. Aiad, F. I. El-Hosiny, and A. Mohsen. Studying the impact of admixtures chemical structure on the rheological properties of silica-fume blended cement pastes using various rheological models. *Ain Shams Eng. J.*, 2021.

55. S. Gupta, P. Krishnan, A. Kashani, and H. W. Kua. Application of biochar from coconut and wood waste to reduce shrinkage and improve physical properties of silica fume-cement mortar. *Constr. Build. Mater.*, 2020.

56. A. Abdullah, M. Taha, M. Rashwan, and M. Fahmy. Efficient use of graphene oxide and silica fume in cement-based composites. *Materials (Basel).*, 2021.

57. X. Ma, T. He, Y. Xu, R. Yang, and Y. Sun. Hydration reaction and compressive strength of small amount of silica fume on cement-fly ash matrix. *Case Stud. Constr. Mater.*, 2022.

58. R. Banar, P. Dashti, A. Zolfagharnasab, A. M. Ramezanianpour, and A. A. Ramezanianpour. A comprehensive comparison between using silica fume in the forms of water slurry or blended cement in mortar/ concrete. *J. Build. Eng.*, 2022.

59. M. M. Al-Zahrani, K. A. Alawi Al-Sodani, M. Maslehuddin, O. S. Baghabra Al-Amoudi, and S. U. Al-Dulaijan. Chloride diffusion models for type V and silica fume cement concretes. *ACI Mater. J.*, 2020.

60. K. A. A. Al-Sodani, M. Maslehuddin, O. S. B. Al-Amoudi, T. A. Saleh, and M. Shameem. Performance of corrosion inhibitors in cracked and uncracked silica fume cement concrete beams. *Eur. J. Environ. Civ. Eng.*, 2020.

61. M. Mejdi, M. Saillio, T. Chaussadent, L. Divet, and A. Tagnit-Hamou. Hydration mechanisms of sewage sludge ashes used as cement replacement. *Cem. Concr. Res.*, 2020.

62. L. Świerczek, B. M. Cieślik, and P. Konieczka. Challenges and opportunities related to the use of sewage sludge ash in cement-based building materials: A review. *Journal of Cleaner Production*. 2021.

63. S. A. L. I. Hassooni and S. Ethaib. Evaluation the effect of reuse sewage sludge and sewage sludge ash on concrete for cement replacement. *J. Eng. Sci. Technol.*, 2020.

64. E. Haustein, A. Kuryłowicz-Cudowska, A. Łuczkiewicz, S. Fudala-Książek, and B. M. Cieślik. Influence of cement replacement with sewage sludge ash (SSA) on the heat of hydration of cement mortar. *Materials (Basel).*, 2022.

65. Z. Chang, G. Long, Y. Xie, and J. L. Zhou. Chemical effect of sewage sludge ash on early-age hydration of cement used as supplementary cementitious material. *Constr. Build. Mater.*, 2022.

66. C. Gu, Y. Ji, Y. Zhang, Y. Yang, J. Liu, and T. Ni. Recycling use of sulfate-rich sewage sludge ash (SR-SSA) in cement-based materials: Assessment on the basic properties, volume deformation and microstructure of SR-SSA blended cement pastes. *J. Clean. Prod.*, 2021.

67. Y. F. Zhou, J. S. Li, J. X. Lu, C. Cheeseman, and C. S. Poon. Recycling incinerated sewage sludge ash (ISSA) as a cementitious binder by lime activation. *J. Clean. Prod.*, 2020, **244**.

68. R. Lumor, L. Abladey, D. Tikoli, A. K. Gand, E. O. Owusu, K. Offei-Nyako, and I. E. Edim. A comparative study of the quality of sandcrete cement blocks and quarry dust cement blocks. *Eng. Solid Mech.*, 2021.

69. K. C. Onyelowe, D. B. Van, L. Dao-Phuc, F. Onyelowe, C. Ikra, C. Ezugwu, A. B. Salahudeen, M. Maduabuchi, J. Obimba-Wogu, K. Ibe, and L. Ihenna. Evaluation of index and compaction properties of lateritic soils treated with quarry dust based geopolymer cement for subgrade purpose. *Epa. J. Silic. Based Compos. Mater.*, 2020.

70. M. H. Elseknidy, A. Salmiaton, I. N. Shafizah, and A. H. Saad. A study on mechanical properties of concrete incorporating aluminum dross, fly ash, and quarry dust. *Sustainability*, 2020.

Chapter 15

Natural Rubber Latex and Its Effects on Some Properties of Bitumen for Cleaner Built Environment

Theresah Osei,[a] Trinity A. Tagbor,[a] Johannes A. M. Awudza,[b] and Mehmet Serkan Kirgiz[c]

[a]*Council for Scientific and Industrial Research-Building and Road Research Institute, Ghana*

[b]*Department of Chemistry, Kwame Nkrumah University of Science and Technology, Ghana*

[c]*Academia of ACMCEN, Beaverton, OR 97075, USA*

nakres42@yahoo.com

In many countries, the main binder for asphalt roads is bitumen. However, there are many cases of early road deterioration resulting sometimes from the effects of temperature on the asphalt binder (bitumen). For this reason, several studies are ongoing to develop bitumen suitable for the climatic conditions in the country. This study examines some physical properties such as softening point temperature, viscosity, penetration temperature, specific gravity and chemical functionalities, peak intensity, and a new functional

Handbook of Innovative Adhesive Technology: Biodegradable and Cost-Effective Binder for Built Environment and Engineering Sciences

Edited by Mehmet Serkan Kirgiz

ISBN 978-981-5129-51-9 (Hardcover), 978-1-003-59892-3 (eBook)

www.jennystanford.com

group of AC 10 grade bitumen modified with natural rubber latex (L) at levels between 1–5% and 10% as well as its effect on aging at 5% and 10% blends. Standard methods of determination of physical properties, such as the Fourier transform infrared spectroscopy (FTIR) test and short-term aging test, were employed. Results indicated that upon the addition of 1% natural rubber latex, all the properties tested decreased. Further increasing the percentage of natural rubber latex decreases the penetration point, but the softening point temperature, specific gravity, and viscosity increase. The spectroscopy analysis result revealed that there was no change in the peak intensities of the blends as compared to the conventional bitumen. However, an enhancement in peak intensity was observed. Aging test results indicated that using 10% latex concentration for bitumen modification for road construction would be highly susceptible to temperature, so the maximum concentration for bitumen modification is 5%. The study has shown that natural rubber could improve some properties of bitumen and could be used for constructing more durable roads.

15.1 Introduction

Bitumen is used mainly as a binder in asphalt pavements for road construction in many developing countries. In recent times, more roads are being constructed for the speedy delivery of goods and services in Ghana. There has also been a general increase in vehicle ownership, resulting in increased pressures from vehicle tires. These pressures have led to a high level of stress exerted on asphalt surfaces, which are exhibited as deformation, surface cracking, fatigue, and reflective cracking, resulting in dangerous road traffic conditions. Ghana depends on imported bitumen, which costs about $250/km in constructing bituminous roads. Since 2013, the country has been importing about 6000 metric tons of bitumen for road construction yearly [1].

Bitumen in asphalt pavement generally softens easily; this situation may become worse due to the relatively high temperatures experienced during some months of the year in a typical tropical country like Ghana. Thus, vehicle tires could easily penetrate the softened asphalt pavement and cause road cracks and severe road potholes [2]. Therefore, there is a need to modify bitumen to increase the softening point temperature as well as decrease the penetration point for better performance and more durable asphalt roads.

Polymer modification of bitumen has generated a lot of research interest worldwide due to the contribution of viscoelastic characteristics to the rheological properties of bitumen. Several studies have reported that the application of polymer materials improves the quality of bitumen by reducing cracks caused by traffic loads, weakness in the cohesion of binders caused by exposure to moisture and temperature, and so on [3].

Polymers normally used as modifiers are styrene butadiene rubber (SBR), styrene butadiene styrene (SBS), polychloroprene, natural rubber, and ethylene vinyl acetate (EVA) [4]. However, natural rubber has a major advantage over other polymers in the formation of emulsions because it can be used in liquid form [5].

The addition of rubber to bitumen impacts the elastic property of bitumen, which increases viscosity and softening point temperature as well as decreases penetration value [6]. The rubber particles swell in the bitumen to form a gel, which increases viscosity. At high temperatures, a rubber-modified asphalt pavement under load stretches without ripping and returns to its original position when the applied load is released. This prevents deformation as well as stripping of asphalt pavement [7]. Rubber also introduces self-healing properties to the asphalt pavement by forming a three-dimensional network around the asphalt binder under load when cracks develop. The binder then softens, and cracks fuse back as the temperature recovers. Rubber in asphalt roads improves the skid resistance of vehicles. The incorporation of natural rubber into bitumen increases the softening point and viscosity, thereby making it less susceptible to temperature [6].

Although Ghana has large plantations of rubber trees, the use of rubber to modify bitumen is relatively uncommon. Meanwhile, it is common to find bituminous roads in a deplorable state due to early deterioration.

This report forms part of results from preliminary studies conducted on locally available natural rubber and AC-10 bitumen to generate scientific data, which may contribute to promoting its use in the construction of bituminous roads.

Table 15.1 describes some of the basic properties of the raw materials used in the study [8, 9].

Table 15.1 Properties of raw materials used in the study

Bitumen	Natural Rubber Latex
• It is chemically inert • Bitumen is thermoplastic • It is insoluble in water • It can adhere to solid surfaces	• Good binding and adhesion properties • Resistance to wear and tear • Great tensile strength • Easily corrodes above 82°C • Resilience to elongation

15.1.1 Objective

The objective of the study was to prepare natural rubber–bitumen composites at different percentages and study their effect on the properties of bitumen. The change in the functional groups of the natural rubber–bitumen composites was also studied using FTIR spectroscopy.

15.2 Materials and Method

15.2.1 Bitumen

The bitumen used for this study is AC-10 grade base bitumen obtained from a local road contractor in Kumasi, Ghana.

15.2.2 Natural Rubber Latex

Natural rubber latex was obtained from the Ghana Rubber Estate Limited, Western Region, Ghana. The rubber was stored in 5% liquid ammonia to prevent coagulation of the latex.

15.2.3 Sample Preparation

The blends were prepared using melt blend techniques. The bitumen was weighed in a stainless steel bucket and heated to a temperature of 160°C on a Sybron Thermolyne HP-A1915B hot plate. At fluid condition, the latex was slowly added at percentages of 1, 2, 3, 4, 5, and 10, while the speed of the mixer was maintained at 120 rpm and the temperature was kept at 160°C. Mixing continued for 1 h to produce homogenous mixtures.

The blends were then transferred to 500 g tin containers, labeled, and stored for further testing. No samples were prepared for 6–9% due to inadequate funding and also because earlier studies reported optimum performance between 4% and 5% [10, 11]. However, a 10% replacement was prepared and studied, the results of which will encourage future studies.

15.2.4 Laboratory Analysis

The following analyses were carried out on the prepared composite:

- **Penetration point:** This test was done according to the test protocol stated in ASTM D5-97.
- **Softening point temperature:** The test was done according to the method described in ASTM D 36-70.
- **Kinematic viscosity:** Measurements were made at 135°C according to the test method stated in ASTM D-2170-10.
- **Specific gravity:** The specific gravity of the bituminous mixture was measured according to the test protocol described in ASTM D 70-76.
- **Short-term aging test:** This test was done using the rotating cylinder aging test using the procedure stated in ASTM D 1754.
- **Flash point:** The flash point test was done according to the test procedure described in ASTM D3143.
- **FTIR test:** FTIR analysis was conducted on the crude bitumen, natural rubber latex, and the blends. The FTIR spectra were measured using an Interspec 200-X. The crude bitumen and the binders were heated to fluid condition, and 0.2 g of each sample was dissolved in 2 mL Analar-grade n-hexane and the test was conducted. All spectra were obtained at wavenumbers ranging from 4000 cm^{-1} to 400 cm^{-1}.

15.2.5 Aging Index

The aging index is calculated using the following equation [12]:

$$\text{VAI} = \frac{n(\text{aged}) - n(\text{unaged})}{n(\text{unaged})}$$

where VAI is the viscosity aging index, *n*(aged) is the viscosity of bitumen after aging, and *n*(unaged) is the viscosity of bitumen before aging.

15.2.5.1 Penetration index and temperature susceptibility

The penetration index and temperature susceptibility are calculated using the following equation [13]:

$$\text{PI} = \frac{1952 - 500 \times \log(\text{Pen25}) - 20 \times \text{SP}}{50 \times \log(\text{Pen25}) - \text{SP} - 120}$$

A = (log PEN @ 25°C - log 800)/(25 – ASTM softening point)

where PI is the penetration index, Pen25 is the penetration at 25°C in a tenth of millimeters, SP is the softening point temperature in degrees Celsius, and A is temperature susceptibility.

15.2.5.2 Specifications by Ghana Highway Authority

Table 15.2 gives the acceptable limit of bitumen for road construction by the Ghana Highway Authority (GHA).

Table 15.2 Specifications for unmodified bitumen by the GHA

Property of Bitumen	Specification
Softening point temperature (°C)	48–56
Viscosity (Cs) at 135°C	≥300
Penetration (dmm) at 25°C, 100 g, 5 s	—
Specific gravity	1.01–1.06
Percentage loss in mass after aging	—

15.3 Results of Analysis

The results of the various tests as well as the spectroscopy test are presented in the following subsections.

15.3.1 Penetration Point

Figure 15.1 represents the results obtained from the penetration test. There was a decrease in penetration value to 4% natural rubber concentration. This increases as the rubber percentage continues to increase.

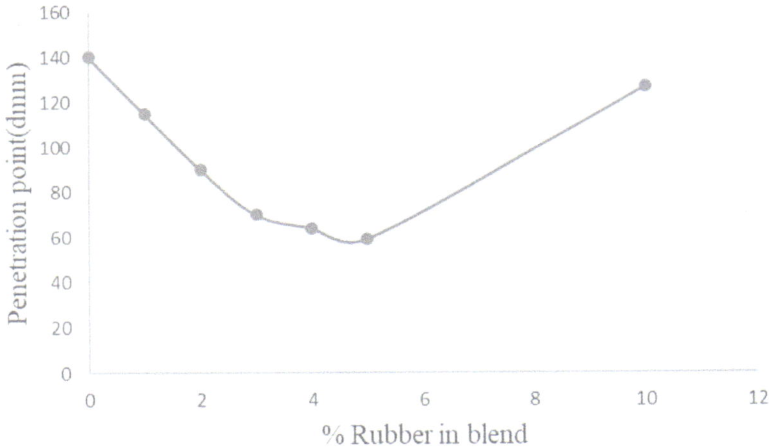

Figure 15.1 Penetration (dmm) value versus latex (%) in modified bitumen.

15.3.2 Softening Point Temperature (°C)

The softening point temperatures at different natural rubber concentrations are presented in Fig. 15.2. The softening point temperature increases along with natural rubber but decreases as the rubber percentage exceeds 4%.

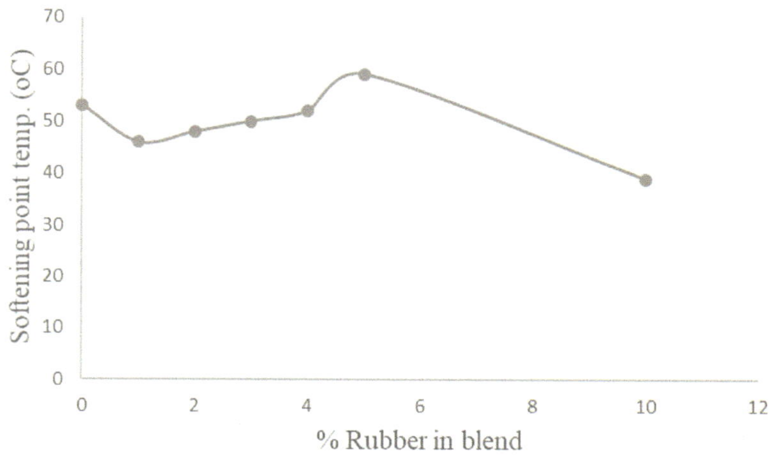

Figure 15.2 Softening point (°C) versus latex (%) in modified bitumen.

15.3.3 Viscosity Analysis

The results of the viscosity test are presented in Fig. 15.3. The viscosity decreased upon the addition of 1% rubber, which then increased as the percentage of natural rubber was increased.

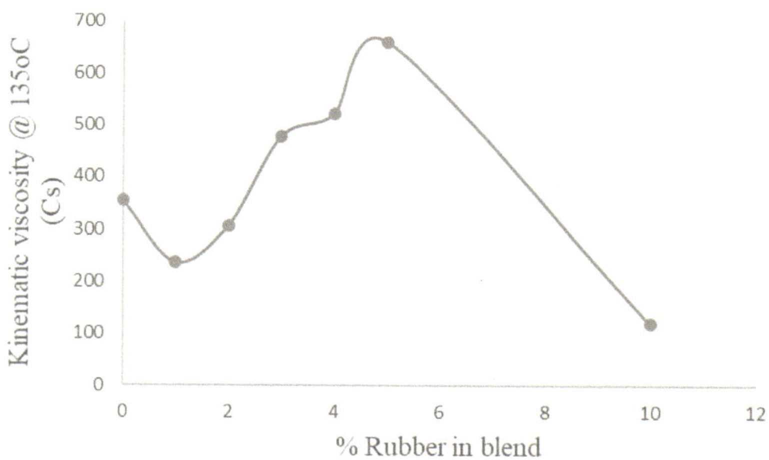

Figure 15.3 Viscosity (Cs) at 135°C versus latex (%) in modified bitumen.

15.3.4 Specific Gravity Test

Figure 15.4 represents the results obtained for the specific gravity test on the blends. The specific gravity recorded an increase as the natural rubber concentration increased but a decrease after 5%.

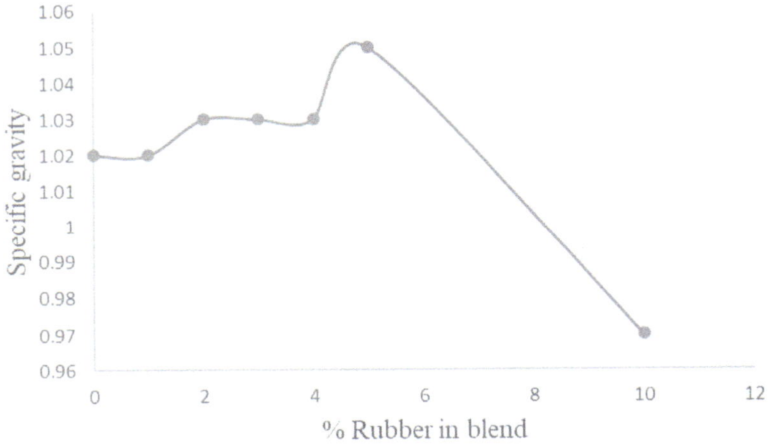

Figure 15.4 Specific gravity versus latex (%) in modified bitumen.

15.3.5 Temperature Susceptibility and Penetration Index

The blend containing 3–5% natural rubber recorded the least temperature susceptibility.

15.3.6 Flash Point, Aging, and Viscosity of Selected Blends

The results obtained for aging, flash point, and viscosity after aging for 5% and 10% natural rubber bitumen blends are presented in Tables 15.3 and 15.4.

Table 15.3 Temperature susceptibility and penetration index

% Blend	Temperature Susceptibility	Penetration Index
0	0.027034	2.756633
1	0.030085	1.979531
2	0.033887	1.134326
3	0.037785	0.383241
4	0.041811	-0.29292
5	0.041526	-0.24802
10	0.028546	2.359427

Table 15.4 Flash point, aging, and viscosity of selected blends

Parameters	Percentage (%) Blend	
—	5%	10%
Flash point (°C)	356.0	377.0
Viscosity before aging (Cst)	660	120.5
Viscosity after aging (Cst)	760.0	1705.0
% Change in mass	0.112	0.135
Aging index	0.152	13.149S

15.3.7 FTIR Analysis

The FRIR result indicates no change in the functional groups but rather recorded an enhancement of peak intensity:

- B – spectrum for crude bitumen
- L – spectrum for natural rubber latex
- BL – spectrum for 5% rubber–bitumen blend

15.4 Discussion

15.4.1 Penetration Point

Penetration depends on the stiffness of the binder. Figure 15.1 depicts a decrease in the penetration value of the bitumen upon the addition of 1% natural rubber latex. This suggests that the penetration value of the modified bitumen is highly dependent on the natural rubber concentrations. The addition of 1%, 2%, 3%, 4%, and 5% natural rubber latex to bitumen decreases the penetration value by 25 dmm, 48 dmm, 70 dmm, 84 dmm, and 85 dmm, respectively, compared to the crude bitumen. A decrease in penetration value with respect to rubber concentration results from the interaction and swelling of rubber particles, which impart elastic properties to the blend, thereby increasing the viscosity. This trend was observed by Cai *et al.* [14] while assessing the performance of SBS latex at different concentrations in bitumen emulsion. Nuha *et al.* [6] also observed a similar trend while analyzing the effect of crumb rubber on the physical and rheological properties of bitumen. The rubber-modified bitumen shows the lowest penetration value between 4% and 5% rubber concentration, which then increases as rubber concentration increases. The increase might have resulted from the coagulation of rubber particles, which then decreased the elastic effect of the blends. Ten percent of latex loading increased the penetration value to 127 dmm, representing an overall change of 13 dmm when compared to unmodified bitumen. This indicates that further increasing the rubber concentration may increase the penetration value, making the blends highly susceptible to temperature and causing deterioration when applied on the road.

15.4.2 Softening Point Temperature

The softening point temperatures of the blends are the reverse of the penetration point when rubber content increases after 1%. The softening point temperature decreased to 46°C at 1% rubber loading and increased with rubber concentration, which then peaked between 4% and 5%. This increase in softening point temperature with respect to rubber content was observed by Visavi and Durga [7] in their research on the effect of natural rubber on bitumen. The increase resulted from the introduction of the elastic property of rubber, causing the blends to be highly viscous and, therefore, increasing the temperature at which the blends soften.

From Fig. 15.2, a steep decrease after 5% rubber concentration can be observed. The decrease may have resulted from increased rubber particles leading to coagulation instead of blending, consequently decreasing the elasticity of the blend.

15.4.3 Viscosity

Figure 15.3 shows the viscosity of modified bitumen blends as the percentage of rubber increases. The addition of 1% latex to bitumen decreases the viscosity to about 30% when compared to unmodified bitumen. This then increases as the rubber concentration increases. This could be ascribed to the fact that the elastic effect at 1% rubber loading was limited and whereas rubber concentration increases, elasticity also increases, therefore resulting in a highly viscous blend. Zhang *et al.* [15] observed a similar development to the above. This may have resulted from the dissolution and dispersion of natural rubber in the bitumen matrix leading to an enhanced viscosity. The decrease in viscosity after 5% rubber levels could be due to an increase in rubber particles, which resulted in coagulation instead of swelling, thereby limiting the elastic effect on the blend.

15.4.4 Specific Gravity

The density of bituminous materials is affected by the chemical composition, temperature, and presence of impurities. Figure 15.4 shows an increase in the specific gravity of blends as rubber concentration increases. The specific gravity increased to 2%, remained constant, and peaked at 5% rubber concentration. The increase in specific gravity may have resulted from the decrease in pores size of bitumen as rubber was introduced, thereby reducing void. The reduction in void would decrease the permeability of moisture and air. Also, the bitumen content in the blends reduces as rubber increases, resulting in an overall increase in specific gravity [16]. This signifies that rubber has a significant effect on bitumen. Therefore, its use in construction may contribute to a decrease in the penetration of moisture and air, thereby a reduction in aging.

15.4.5 Temperature Susceptibility and Penetration Index

Table 15.2 represents the calculated temperature susceptibility and penetration index of the rubber bitumen blends and the conventional bitumen. The table depicts an increase in temperature susceptibility as rubber levels increase, which was within the range of 0.027 to 0.042. The penetration index of modified bitumen was within the ranges of −0.2 and +2.4, which are lower than +2.7 for unmodified bitumen. The penetration index should be within the range of −1 and +1 to make it suitable for pavement construction [17]. This indicates that the blends containing 3%, 4%, and 5% natural rubber could be suitable for pavement construction. Blends containing 1%, 2%, and 10% of natural rubber did not meet the standards stated in EN 12591:2009 [18]. Therefore, they may not be suitable for pavement construction.

15.4.6 Flash Point, Aging, and Viscosity of Selected Blends

Table 15.3 shows the viscosity after aging, flash point, and change in mass after aging. It is observed that the viscosity of the 10% blend before aging was 120 Cst as compared to 1705 Cst after aging, giving an aging index of 14.208. This suggests that the modification of bitumen using 10% natural rubber latex may be highly susceptible to cracks under loading. The high flash point of the blend containing 10% natural rubber suggests an increase in powdered components in the blend. A viscosity increase of 15% after aging with a loss in mass of 0.112% for the blend with 5% natural rubber indicates higher storage stability than the blend containing 10% natural rubber.

15.4.7 FTIR Analysis

From Fig. 15.5, it was observed that the peak displayed at 1639 cm^{-1} corresponding to the sp2 C=C or C=O stretching vibration found on the rubber spectrum could not be seen on the spectrum of modified bitumen. The most intense peak at 3369 cm^{-1}, which corresponds to the stretching vibration of N–H from ammonia, was no longer discernible after blending. This suggests that the incorporation of rubber does not affect the functional structure of the blend. This indicates that the addition of rubber does not influence the functional structure of the blend.

However, there was a shift to higher wavenumber of blends as compared to conventional bitumen. This resulted from the inhibition of hydrogen bonding due to the evaporation of water molecules during blending. There was also no change in the peak intensity of the spectrum of the blend (BL) if compared to the spectrum of crude bitumen (B). This indicates that blending proceeded physically.

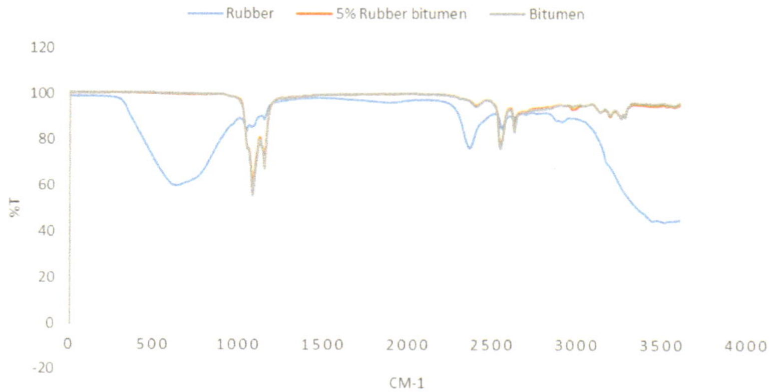

Figure 15.5 FTIR spectroscopy analysis of conventional bitumen (B), latex (L), and 5% rubber–bitumen blend (BL).

15.5 Conclusion

The study analyzed the change in properties of bitumen when natural rubber latex was added at percentages of 1, 2, 3, 4, 5, and 10. The properties tested are penetration point, viscosity, softening point, specific gravity, and change in mass after aging.

FTIR spectroscopy was used to determine changes in the functional groups of blends. The results of this study showed a general decrease in all physical properties tested on bitumen when 1% rubber was added. As the percentage of rubber increases, all the properties increase, except the penetration point. There was no change in the functional groups of the blends when compared to unmodified bitumen. However, there was a shift in the IR spectrum to a higher wavenumber. Short-term aging test results indicate that the addition of 5% natural rubber latex to bitumen showed significant improvement in the properties. The calculated penetration index also showed a significant change in the temperature susceptibility of blends. The study has revealed that the addition of 4–5% natural rubber latex to bitumen could serve as an alternative binder for asphalt pavement.

References

1. UN Statistics Division. 2015. www. kneoma.com.
2. Khodary F, Abd El-sadek MS, and El-Sheshtawy HS. Mechanical properties of modified asphalt concrete mixtures using $Ca(OH)_2$ nanoparticles. *Int J Civ Eng Technol* 2014; **5**(5): 61–68.
3. Ali SA, Yusof I, Hermadi M, Alfergani MBS, and Sinusi AA. Pavement performance with carbon black and natural rubber (Latex). *Int J Eng Technol* 2013; **2**(3): 124–129.
4. Zhu J, Birgisson B, and Kringgos N. Postprint polymer modification of bitumen: Advances and challenges. *Eur Polym J* 2014; **54**: 18–38. http://dx.doi.org/10.1016/j.eurpolymj.2014.02.005.
5. Forbes A, Haverkamp RG, Robertson T, Bryant J, and Bearsley S. Studies of the microstructure of polymer-modified bitumen emulsions using confocal laser scanning microscopy. *J Microsc* 2001; **204**(Pt 3): 252–257. http://dx.doi.org/10.1046/j.1365-2818.2001.00955.x.
6. Nuha SM, Asim HA, Mohamed RK, and Mahrez A. Effect of crumb rubber concentration on the physical and rheological properties of rubberised bitumen binders. *Int J Phys Sci* **6**(4): 684–690.
7. Swetha DV. Effect of natural rubber on the properties of bitumen and bituminous mixes. *Int J Civ Eng Technol* 2014; **5**(10): 9–21.
8. Johnson S. *Latex Specification and Properties*, available at www.sciencing.com/latex-specification.
9. Haseed J. *Properties of Bitumen and Bituminous Material*, available at www.aboutcivil.org/bitumen properties.
10. Abdulrahman S, Hainin MR, Idham MK, Hassan NA, Warid MNM, Yaacob H, Azman M, and Puan OC. Physical properties of warm cup lump modified bitumen. *IOP Conf. Series Material Sci Eng.* 2019; **527**(1): 012048. http://dx.doi.org/10.1088/1757-899X/527/1/012048.
11. Krishnapriya MG. Performance evaluation of natural rubber modified bitumen mixes. *J Civil Environ Water Res Infrastruct Eng* 2015; **5**(1): 1–12.
12. Abutalib N, Elham HF, Aflaki S, and Abu-Lebdeh TM. Investigating effects of application of silica fume to reduce asphalt oxidative aging. *Am J Eng Appl Sci* 2015; **8**(1): 176–184.
13. Whiteoak D and Read J. *The Shell Bitumen Handbook*, 5th edn, London 2003, pp. 43–48.

14. Cai HM, Wang T, Zhang JY, Zhang YZ, Wang T, and Zhang JY. Preparation of an SBS latex-modified emulsion and performance assessment. *Petrol Sci Technol* 2010; **28**(10): 987–996. http://dx.doi.org/10.1080/10916460902939436.
15. Zhang J, Wang J, Wu Y, Sun W, and Wang Y. Thermal behaviour and improved properties of SBR and SBR/natural bitumen modified bitumens. *Iran Polym J* 2009; **18**(6): 465–478.
16. Blazejowski K, Olszacki J, and Peciakowski H. *Bitumen Handbook*, 2014.
17. Abir R. Waste rubber bitumen modifier. *Int J Eng Res Gen Sci* 2015; **3**(5): 109–119.
18. EN 12591:2009. *Bitumen and Bituminous Binders Specification for Paving Grade Bitumen.*

Chapter 16

Analysis of E-Waste Containing Carbon and Other Components as Composite Material for a Cleaner Built Environment and Realizing UN Sustainable Development Goals

M. Vishnu Priyan,[a] R. Annadurai,[a] Mehmet Serkan Kirgiz,[b] André Gustavo de Sousa Galdino,[c] Raul D. S. G. Campilho,[d] Roberto Alonso González-Lezcano,[e] and Konstantinos G. Kolovos[f]

[a]*Department of Civil Engineering, SRM Institute of Science and Technology, Kattankulathur, Chengalpattu, Tamil Nadu, 603203, India*
[b]*Academia of ACMCEN, Beaverton, OR 97075, USA*
[c]*Federal Institute of Education, Sciences and Technology of Espírito Santo, Av. Vitória, 1729, Jucutuquara, Vitória, ES, 29040-780, BR*
[d]*Departamento de Engenharia Mecânica, Instituto Superior de Engenharia do Porto, Instituto Politécnico do Porto, R. Dr. António Bernardino de Almeida, 431, 4200-072 Porto, Portugal*
[e]*Architecture and Design Department, Escuela Politécnica Superior, Universidad CEU San Pablo, CEU Universities, 28040 Madrid, Spain*
[f]*Department of Military Sciences, Division of Physical Sciences and Applications, Hellenic Army Academy, Vari, Greece*
nakres42@yahoo.com

Handbook of Innovative Adhesive Technology: Biodegradable and Cost-Effective Binder for Built Environment and Engineering Sciences
Edited by Mehmet Serkan Kirgiz

ISBN 978-981-5129-51-9 (Hardcover), 978-1-003-59892-3 (eBook)
www.jennystanford.com

The purpose of the chapter is to review manuscripts on the use of E-waste in composites in terms of a cleaner built environment and realizing the Sustainable Development Goals of the United Nations. For that aim, we collected manuscripts that used E-waste as an aggregate stack in composite mixing. The chapter evaluates the composite containing E-waste as an aggregate stack in terms of specific gravity, water absorption, and compressive testing. Also, comprehensive results were obtained. In conclusion, the chapter represents that E-waste can be used as an aggregate stack in composite to advance a cleaner built environment and realize the Sustainable Development Goals of the United Nations.

16.1 Introduction

The composite is the main material used by the construction industry for long-lasting buildings. Swift construction and an increase in population have led to the demand for building additional structures, which has increased the need for the main ingredients of composite, such as aggregates, cement, and water, resulting in a scarcity of natural resources. Since natural aggregates are not a sustainable resource, these need to be replaced in the composite by materials such as human-made waste. In addition, making construction materials friendly to the environment is critical for saving natural resources. Moreover, the composite's pros and cons are, respectively, its strength in compression and lack of tension. The particles of cement and aggregate result in higher or lower compression, which is the primary problem in composites manufactured from unnatural resources [1, 2].

E-waste is surplus as broken electrical and electronic items are dumped carelessly. The electronic sector has generated a stream of electronic waste, which contains oxides of carbon, titanium, aluminum, copper, lithium, nickel, silicon, and zinc. Also, E-waste contamination harms the environment. As a result, the most appropriate course of action is incorporating E-waste with composite. Recycling a significant percentage of E-waste can be achieved by the composite industry, which would overcome the waste disposal problem. Also, some options need to be considered, especially regarding the recycling of material units [3].

16.1.1 Sources of E-waste and Its Stance in the World

E-waste can be generated from various sources, such as electronic appliances used by people. Household appliances are the most common source of E-waste. Of all appliances, personal computers, DVD players, laptops, television sets, cell phones, and other devices are the most important examples. People believe that E-waste consists exclusively of information technology items; however, when electrical and other household equipment are thrown away, they make E-waste a possible source of contamination. Such items include washing machines, vacuum cleaners, toasters, drying machines, refrigerators, air conditioners, and other types of electronics.

Another type of electrical item is a printed circuit board (PCB). Regularly, components of PCB consist of about 40% metals, 30% polymers, and 30% ceramics [4, 5]. They can be divided into three categories: organic, metal, and ceramic. The organic group in PCB includes flame-retardant polymer and paper. The metallic group contains primary metals such as copper, iron, and aluminum and also includes expensive metals such as tantalum, gallium, silver, carbon, and palladium. PCB ceramic is composed of oxides of silicon and aluminum. PCB particularly contains toxic elements such as chromium, lead, beryllium, mercury, and cadmium. Additionally, oxides of alkaline earth, mica, and barium are included. Metals can be reused, while nonmetals can be repurposed [6]. Almost 70% of the predicted heavy metals originate from electronics abandoned in the United States of America [7]. In India, Maharashtra has the highest percentage of E-waste, followed by Tamil Nadu and Andhra Pradesh [8–13]. According to the United Nations Environment Program, India's E-waste production will increase by 500% over the next 10 years [14]. The United States is the world leader in disposing of approximately 11.7 million tons of E-waste annually. China is behind the United States, with 6.1 million tons of E-waste. In most cases, only 15–20% of E-waste is recycled, and the rest is disposed of in landfills and incinerated [15].

16.2 Materials and Methods

Papers related to the use of E-waste in composites have been evaluated as materials and methods in this chapter to stress the significance of this study.

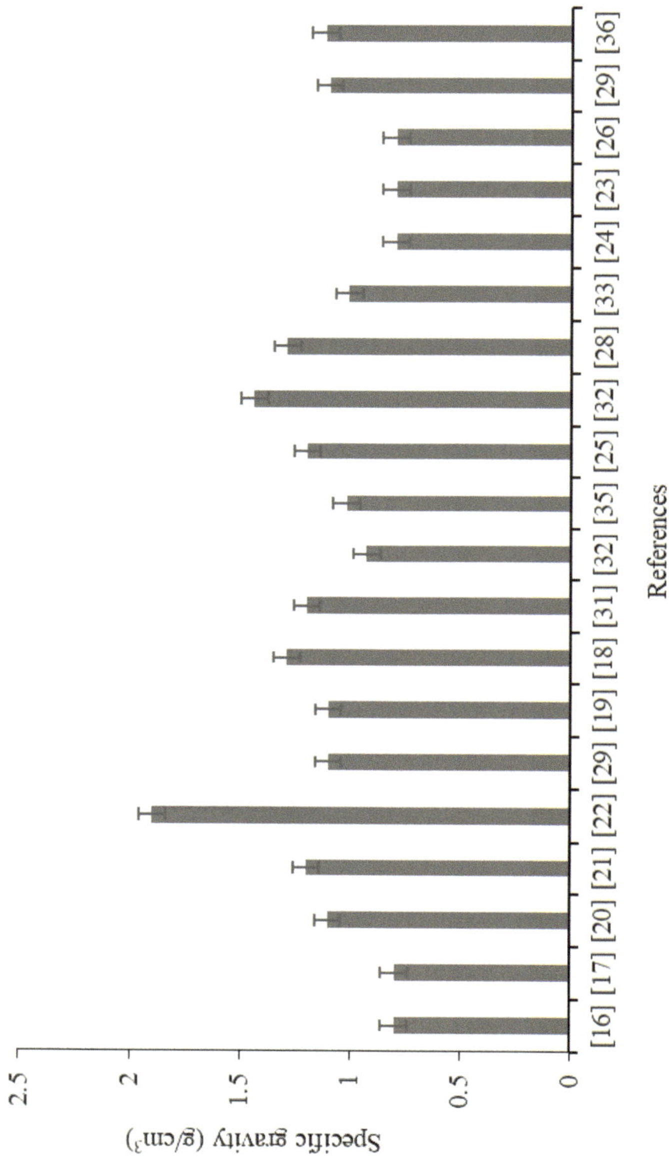

Figure 16.1 Specific gravity achieved by various authors [16–36].

16.3 Results and Discussion

16.3.1 Specific Gravity

Figure 16.1 shows the specific gravity for the materials used. Specific gravity deals with the ratio of material density to water density at room temperature.

The material volume of E-waste is the same as that of the other natural stack of fine and coarse aggregates, but the mass varies for every material added to the composite. In this chapter, the specific gravity values of the different kinds of E-waste added to the composite specimen have been inclusively graphed in Fig. 16.1 related to various researchers. The research done by Suchithra *et al.* [20] has the best specific gravity for a composite containing E-waste.

16.3.2 Water Absorption

Figure 16.2 shows the results of water absorption reviewed and documented by various authors around the globe.

The water absorption capacity has a significant role in the hydration process of composite specimens. The natural aggregate stack consumes more water than the synthetic aggregate stack. The lower water absorption capacity of the mixing ingredient of the composite decreases the quantity of mixing water inserted for the hydration process. Many researchers have tested the water absorption capacity of E-waste, and they found that the water absorption capacity of E-waste is very low when compared to the other coarse and fine aggregate stacks in nature [28, 32, 33]. Absorption testing can be used to determine the effective porosity of the material. Kumar *et al.* [28] showed the possibility of manufacturing composites using various percentages of E-waste, such as 0%, 10%, 20%, 30%, 40%, and 50% by volume, as the partial alternative for natural coarse aggregates. The percentage of water absorbed by concrete containing E-waste is more significant than that of the composite without E-waste. Although the integrated E-wastes increased the composite's water absorption capacity, the values are within the acceptable range.

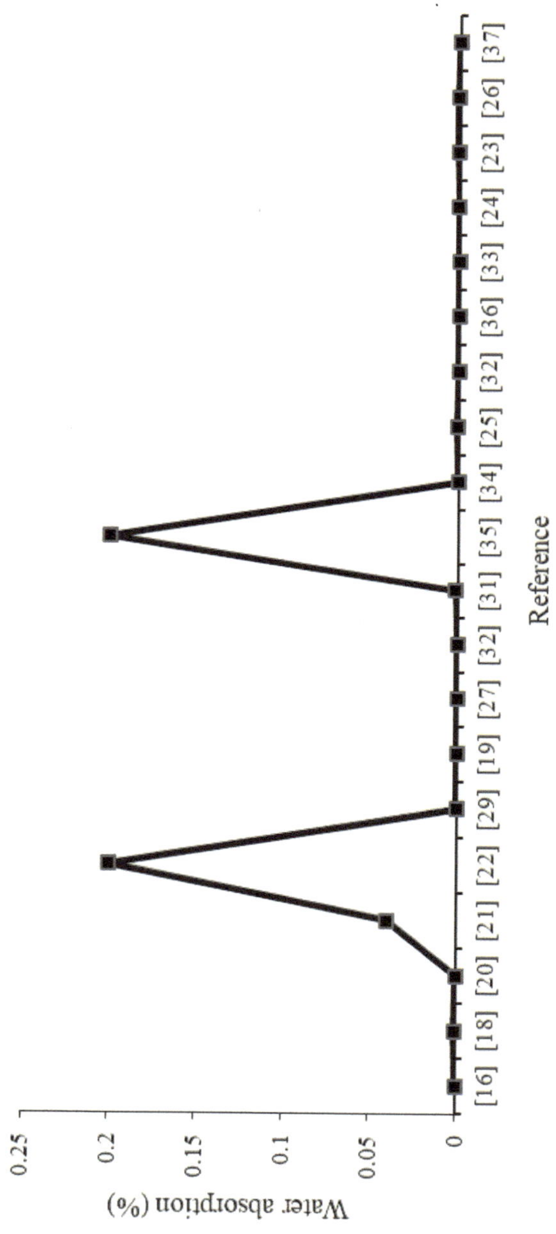

Figure 16.2 Water absorption found in various published works [16, 18–29, 31–37].

16.3.3 Compressive Strength

Figure 16.3 shows the compressive strengths of composites containing E-waste for various strength classes. For all composites reviewed, the compressive strength of the composite made by replacing the natural aggregate stack with E-waste was more or less equal to or greater than the strength of conventional composites on the 28th day of curing [35].

However, a few authors reported that the strength of the composite containing E-waste decreased gradually compared to the control composite. However, a majority of researchers have achieved a higher strength than conventional composites. The strength of the composite made with E-waste is 20% higher than that of the control composite [34, 43]. When E-plastic is substituted for coarse aggregates to an extent of 20%, it does not reduce the composite's compressive strength. Bharani *et al.* [44] found that the composite mixed with E-waste yielded a higher strength by replacing 20% of its coarse aggregates. Shinu and Needhidasan [16] found that E-waste reduced the composite strength due to its weak bond with cement paste and increased the matrix porosity, reducing density and strength [16]. Gavhane *et al.* [45] recognized the possibility of using E-waste as an alternative to natural coarse aggregates, with up to 10% replacement by volume. A lower difference in strength was observed when replaced with 10% E-waste after 28 days of curing [45]. Prasanna and Rao [21] undertook a study by partially replacing the mixing ingredients of composite with E-waste. After 28 days of curing, the strength of the control specimen was reduced by about 20% due to the addition of E-waste, and another 10% due to the addition of fly ash.

According to Arora and Dave [46], E-waste can be used as fine aggregate in mortar on a limited basis. By replacing 4% of E-waste with fine aggregate, a good strength gain could be achieved, and replacing 0–24% of aggregate in concrete with E-waste noticeably increases the strength. A decrease in strength has been detected at every substitution level. It has been observed that the most substantial decline in strength occurred at 20% replacement of coarse aggregates with E-waste [47]. The compaction strength also improved with a 10% fly ash mineral admixture [29, 31, 48]. A

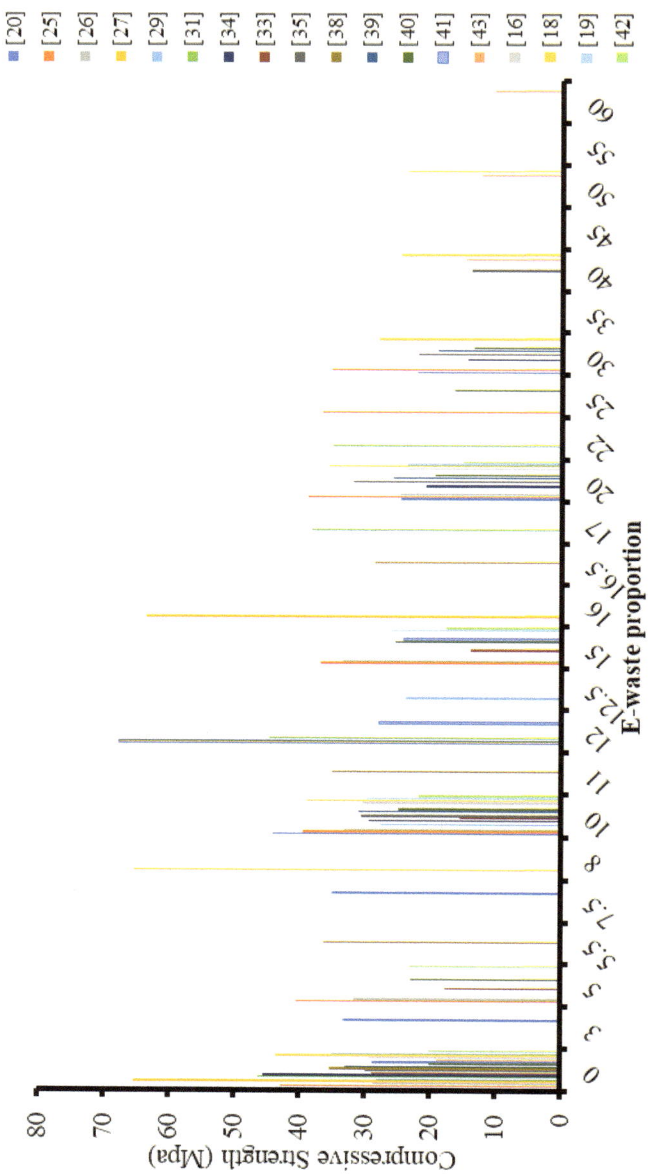

Figure 16.3 Effect of E-waste on compressive strength after 28 days [16, 18–20, 25–27, 29, 31, 33–35, 38–42].

compressive strength higher than that of the reference composite has been achieved with a 12% replacement of natural aggregates with E-waste. Kala and Nagajothi [12, 49] reported that by adding E-fiber waste up to 2.5% of composite volume, the compressive strength was almost two times higher than that of the conventional mix. The addition of E-fiber waste increased the strength of the composite. According to Chen *et al.* [50], the composite with 40% replacement of coarse aggregates demonstrates 17%, 27%, and 43% higher compressive strengths than the control composite after 28, 91, and 365 consecutive days, respectively. E-waste particles have also been observed to possess crack resistance when incorporated with the composite. E-waste particles resist composite cracks [38–41].

16.3.4 Split Tensile Strength

Figure 16.4 shows the effect of E-waste on tensile strength after 28 days. E-waste concrete increases the split tensile strength by 20% compared to the control concrete. If E-plastic is substituted for coarse aggregates to an extent of 20% substitution, it does not reduce the concrete's split tensile strength [26]. Any substitution above 20% reduces concrete strength. Split tensile strength is greater when 12.5% coarse aggregate is substituted with E-waste [23, 42, 43]. The flexural strength of E-waste concrete decreases with the increasing E-waste, whereas the tensile strength increases with increasing electronic plastic in concrete. Muthadhi *et al.* [44] found that with up to 10% E-waste replacement in conventional concrete, the tensile strength increased marginally after 28 days of curing. Alagusankareswari *et al.* [45] used E-waste as a replacement for fine aggregates and found that E10, E20, and E30 mixed media became firm at a slower rate than E-waste.

The split tensile strength of E-waste concrete is slightly lower than that of regular concrete. Suchithra *et al.* [22] investigated the use of E-waste in M20 grade concrete to the extent of up to 20% replacement of coarse aggregates. It has been evident that E-waste has a more significant impact on flexural strength than split tensile strength. Lakshmi and Nagan [31] reported that E-waste was replaced from 0% to 24% for coarse aggregates during the concrete replacement trial, and the substantial concrete at 20% replacement had the most strength reduction. Additionally, 10% fly ash was mixed with concrete as a mineral admixture, resulting in 50% greater split tensile strength than the reference concrete [46–49]. The strength of the concrete was better, with up to 12% replacement of natural aggregates. Many researchers have tested conventional and E-waste concrete specimens for split tensile strength. The results show that conventional concrete has less strength when compared to all other E-waste concrete. Jose and Sangeetha [50] have investigated the split tensile strength of concrete with E-fiber. The results indicate that specimens with 15% coarse aggregate replaced with E-fiber had a 2.26% increase in tensile strength compared to the standard concrete specimen. The controlled specimens failed to be brittle, but the specimens reinforced with E-plastic and E-fiber had enormous elastic deformation before failure. As a result, it frequently exhibited a ductile failure mode. The inclusion of E-fiber resulted in a significant increase in tensile strength compared to conventional specimens.

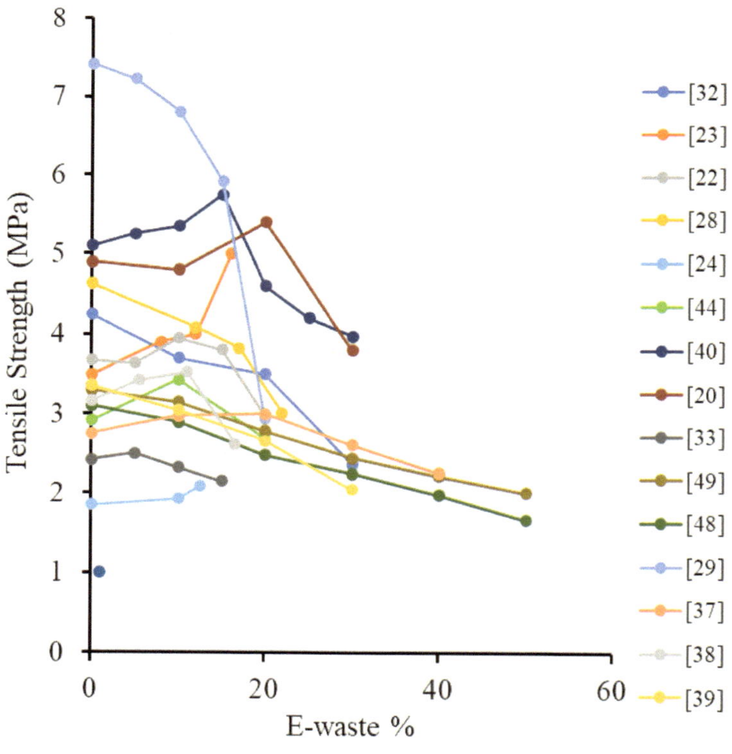

Figure 16.4 Effect of E-waste on tensile strength after 28 days.

16.3.5 Flexural Strength

Figure 16.5 shows the effect of E-waste on flexural strength after 28 days of aging. Santhanam and Anbuarasu [23] studied flexural strength using M-sand and E-waste replaced coarse aggregate (0–12.5%) in M20 grade concrete. The flexural strength of concrete increases significantly when 10% of coarse aggregate is replaced with E-waste. With an increase in E-waste, the concrete has a reduced flexural strength [51]. Bharani *et al.* [27] found that concrete mixed with E-waste yielded a slight increase in flexural strength when 20% of the coarse aggregate was replaced.

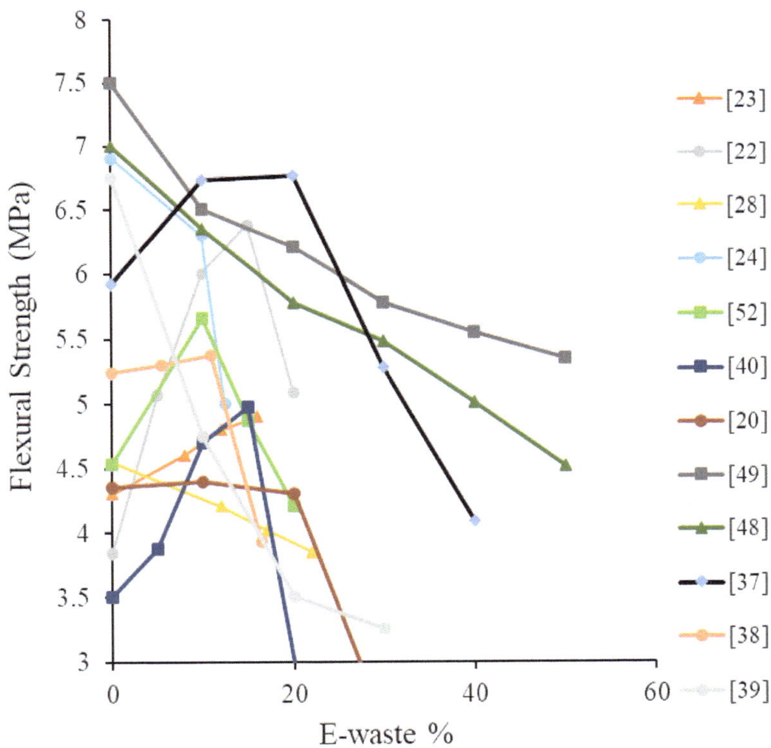

Figure 16.5 Effect of E-waste on flexural strength after 28 days of aging.

Rathore and Rawat [42] reported that 15% E-waste substitution increased the flexural strength of concrete when compared with conventional concrete. From the aforementioned studies, it was found that 15% is the optimum replacement percentage for E-waste in construction. Figure 16.5 demonstrates E-waste concrete's flexural strength for various grades with various percentages of E-waste. The addition of E-waste reduces compressive strength but improves flexural strength at later ages [52]. In M20 grade concrete, Suchithra *et al.* [22] replaced aggregate with E-waste by up to 20%. E-waste significantly increases flexural strength by up to 15%. It seems that E-waste has a more significant impact on flexural strength [46]. In their experimental investigation, the researchers employed a combination of electronic waste as a partial substitute for coarse materials in concrete. Semiyaga *et al.* [53] evaluated the benchmark characteristics of E-waste to include them in concrete as a substitution.

When 5–20% of coarse aggregate is replaced with E-plastic, a reduction in compressive, splitting tensile, and flexural strengths of 42.35%, 36.33%, and 40.98%, respectively, is observed with 20% E-wastes in concrete at the highest.

16.4 Conclusion

This chapter provides an analytical and exhaustive analysis of the feasible recycling methods for reusing E-waste as a building material and its alternatives. An investigation was conducted to assess the effects of E-waste on the properties of mortar, composite, and precast components. In this article, we described the numerous characteristics of composites that incorporate E-waste. The following inferences are derived based on this investigation:

- The inclusion of E-waste in composite has resulted in a lightweight composite. E-waste composite can be used to construct nonstructural elements. Due to its higher durability against water absorption, it can build sewage treatment plants, chemical plants, seashore buildings, and marine structures.
- The compressive strength of fine aggregate replaced with E-waste composite significantly increased after 5–20% of replacement. However, beyond that, the compressive strength decreases gradually. Compressive, flexural, and split tensile strengths of concrete reinforced with E-waste fiber all increase over those of conventional concrete. Overdoing it with the concrete's fiber content can reduce its mechanical properties.

- In the composite, replacing coarse aggregate by 10–25% with E-waste will marginally increase the compressive strength. As the proportion of substation increases, the workability of the E-waste composite decreases but can be significantly overcome by adding a superplasticizer or fly ash, compared to other materials. With a greater proportion of E-waste, the self-weight of the composite decreases significantly.

References

1. Vishnupriyan, M., Annadurai, R., and Bahrami, A. Experimental investigation of mechanical and durability properties of waste printed circuit board (WPCB) reinforced concrete and validation using response surface methodology. *Front Mater* **10**, 1189274, doi:10.3389/FMATS.2023.1189274.
2. Mirza, J., Huseien, G., and K.S.-J.A. Reduction in ecology, environment, economy and energy in concrete industry using waste materials. 2020, onlinejournal.org.uk.
3. Akinseye, V.O. Electronic waste components in developing countries: Harmless substances or potential carcinogen. *Ann Res Rev Biol* 2013, **3**, 131–147.
4. Das, A., Ari, V., and Mehrotra, S.P. A novel flowsheet for the recovery of metal values from waste printed circuit boards. *Res Conserv Recycl* 2009, **53**, 464–469, doi:10.1016/j.resconrec.2009.03.008.
5. Sodha, A.B., Qureshi, S.A., Khatri, B.R., Tipre, D.R., and Dave, S.R. Enhancement in iron oxidation and multi-metal extraction from waste television printed circuit boards by iron oxidizing leptospirillum feriphillum isolated from coal sample. *Waste Biomass Valorization* 2019, **10**, 671–680. doi:10.1007/s12649-017-0082-z.
6. Donadkar, U., Solanke, M., and Shrikant, S. Review of E-waste material used in making of concrete. *Int J Sci Technol Eng* 2016, **2**, 66–69.
7. Pariatamby, A., and Victor, D. Policy trends of e-waste management in Asia. *J Mater Cycles Waste Manag* 2012, **15**, 411–419. doi:10.1007/s10163-013-0136-7.
8. Gavhane, A., Soni, S., Sutar, D., and Patil, M.P. Utilisation of e-plastic waste in concrete. *Int J Eng Res Technol* 2016, **5** (2), 594–601.
9. Pandiyarajan, M. Behavior of e-waste plastics in concrete: Experimental study on e-waste concrete and comparing with conventional concrete, 2015. doi:10.13140/RG.2.2.31799.24489.

10. Vishnupriyan, M., and Annadurai, R. Investigation of the effect of substituting conventional fine aggregate with PCB powder on concrete strength using artificial neural network. *Asian J Civ Eng* 2023, **1**, 1–9. doi:10.1007/S42107-023-00700-7.

11. Vishnupriyan, M., and Annadurai, R. A study on the macro-properties of PCB fiber-reinforced concrete from recycled electronic waste and validation of results using RSM and ANN. *Asian J Civ Eng* 2023, **1**, 1–14. doi:10.1007/S42107-023-00595-4.

12. Marimuthu, V., and Ramasamy, A. Investigation of the mechanical properties of M40-grade concrete with PCB fiber from recycled electronic waste. *J Hazard Toxic Radioact Waste* 2023, **27**, 4022034, doi:10.1061/(ASCE)HZ.2153-5515.0000725.

13. Marimuthu, V., and Annadurai, R. Experimental investigation on the mechanical properties of printed circuit board (PCBs) fibre reinforced concrete, in *Proceedings of International Conference on Novel Innovations and Sustainable Development in Civil Engineering*, March 31 and April 1, 2022.

14. Mohd, S., and Kaushal, V. E-waste management in India: Current practices and challenges, 2018.

15. Fu, J., Zhang, H., Zhang, A., and Jiang, G. E-waste recycling in China: A challenging field. *Environ Sci Technol* 2018, **52**, 6727–6728.

16. Mary Treasa Shinu, N.M., and Needhidasan, S. An experimental study of replacing conventional coarse aggregate with e-waste plastic for M40 grade concrete using river sand, in *Proceedings of the Materials Today: Proceedings*, Elsevier Ltd, 2020, **22**, pp. 633–638.

17. Needhidasan, S., Samuel, M., and Chidambaram, R. Electronic waste: An emerging threat to the environment of urban India. *J Environ Health Sci Eng* 2014, **12**, 36. doi:10.1186/2052-336X-12-36.

18. Manjunath, B.T.A. Partial replacement of e-plastic waste as coarse-aggregate in concrete. *Procedia Environ Sci* 2016, **35**, 731–739. doi:10.1016/j.proenv.2016.07.079.

19. Ahirwar, S., Malviya, P., Patidar, V., and Singh, V.K. An experimental study on concrete by using E-waste as partial replacement for course aggregate, academia.edu 2016, 3.

20. Suchithra, S., Manoj, K., and Indu, V.S. Study on replacement of coarse aggregate by E-waste in concrete. *Int J Eng Res Technol* (IJERT) 2015, **3**, 266–270.

21. Prasanna, P.K., and Rao, M.K. Strength variations in concrete by using E-waste as coarse aggregate. *Int J Edu Appl Res* 2014, **4**.

22. Rathore, V., and Rawat, A. Effective utilization of electronic waste in concrete mixture as a partial replacement to coarse aggregates, in *Proceedings of the AIP Conference Proceedings*, American Institute of Physics Inc., September 25, 2019, 2158.

23. Senthil Kumar, K., and Baskar, K. Recycling of E-plastic waste as a construction material in developing countries. *J Mater Cycles Waste Manag* 2015, **17**, 718–724. doi:10.1007/s10163-014-0303-5.

24. Ling, T.C., and Poon, C.S. Utilization of recycled glass derived from cathode ray tube glass as fine aggregate in cement mortar. *J Hazard Mater* 2011, **192**, 451–456. doi:10.1016/J.JHAZMAT.2011.05.019.

25. Sabău, M., and Vargas, J.R. Use of E-plastic waste in concrete as a partial replacement of coarse mineral aggregate. *Comput Concr* 2018, **21**, 377–384. doi:10.12989/cac.2018.21.4.377.

26. Muthadhi, A., Basid, A.M., Madivarma, R., Satheesh Kumar, J., Raghuvarman, R., and Prof, A. Experimental investigations on concrete with e-plastic waste. *Int J Sci Eng Res* 2017, **8**.

27. Manatkar, A.P. Use of non-metallic E-waste as a coarse aggregate in a concrete. *Int J Res Eng Technol* 2015, **04**, 242–246. doi:10.15623/ijret.2015.0403040.

28. Kumar, A., Holuszko, M., and Espinosa, D.C.R. E-waste: An overview on generation, collection, legislation and recycling practices. *Resour Conserv Recycl* 2017, **122**, 32–42. doi:10.1016/J.RESCONREC.2017.01.018.

29. Selvam, P.N., and Gopal Krishna, G.V.T. Recycle of E-waste in concrete. *Int J Sci Res* 2013.

30. Senthil Kumar, K., and Baskar, K. Development of ecofriendly concrete incorporating recycled high-impact polystyrene from hazardous electronic waste. *J Hazard Toxic Radioact Waste* 2015, **19**, 04014042. doi:10.1061/(asce)hz.2153-5515.0000265.

31. Akram, A., Sasidhar, C., and Mehraj Pasha, K. E-waste management by utilization of E-plastics in concrete mixture as coarse aggregate replacement. *Int J Innov Res Sci Eng Technol* 2007, **3297**. doi:10.15680/IJIRSET.2015.0407008.

32. Santhanam, N., Anbuarasu, G., and Ramesh, B. Study on partial replacement of coarse aggregate using PCB trimmers with fly ash and without fly ash, in *Proceedings of the Materials Today: Proceedings*, Elsevier Ltd, 2020, **43**, pp. 1931–1937.

33. Santhanam, N., and Anbuarasu, G. Experimental study on high strength concrete (M60) with reused E-waste plastics. *Mater Today Proc* 2020, **22**, 919–925. doi:10.1016/J.MATPR.2019.11.107.

34. Arivalagan, S. Experimental study on the properties of green concrete by replacement of E-plastic waste as aggregate, in *Proceedings of the Procedia Computer Science*, Elsevier B.V., 2020, **172**, pp. 985–990.

35. Kumar, A. Utilization of e-waste in concrete by partial replacement of coarse aggregate, 2018.

36. Rohini, I., and Padmapriya, R. Effect of bacteria subtilis on E-waste concrete, in *Proceedings of the Materials Today*: *Proceedings*, Elsevier Ltd, 2020, **42**, pp. 465–474.

37. Needhidasan, S., Samuel, M., and Chidambaram, R. Electronic waste: An emerging threat to the environment of urban India. 2014, doi:10.1186/2052-336X-12-36.

38. Mane, K.M., Nadgouda, P.A., and Joshi, A.M. An experimental study on properties of concrete produced with m-sand and e-sand, in *Proceedings of the Materials Today*: *Proceedings*, Elsevier Ltd, 2020, **38**, pp. 2590–2595.

39. Suleman, S., and Needhidasan, S. Utilization of manufactured sand as fine aggregates in electronic plastic waste concrete of M30 Mix, in *Proceedings of the Materials Today Proceedings*, Elsevier Ltd, 2020, **33**, pp. 1192–1197.

40. Asha, P. Study of strength characteristics of concrete with E-waste. The *International Journal of Earth Sciences and Engineering* 2015, **8**(1), 237–241. ISSN 0974-5904.

41. Balasubramanian, B., Gopala Krishna, G.V.T., and Saraswathy, V. Investigation on partial replacement of coarse aggregate using e-waste in concrete. *Int J Earth Sciences Eng* 2016, **9**, 285–288.

42. Senthil Kumar, K., and Baskar, K. Effect of temperature and thermal shock on concrete containing hazardous electronic waste. *J Hazard Toxic Radioact Waste* 2018, **22**, 04017028. doi:10.1061/(asce)hz.2153-5515.0000387.

43. Prashant, S., and Kumar, M. Effect of partial replacement of coarse aggregates with E-waste on strength properties of concrete, in *Lecture Notes in Civil Engineering*, Springer, 2019, **25**, pp. 555–560.

44. Bharani, S., Rameshkumar, G., Manikandan, J., Balayogi, T., Gokul, M., and Bhuvanesh, D.C. Experimental investigation on partial replacement of steel slag and E-waste as fine and coarse aggregate, in *Proceedings*

of the Materials Today: Proceedings, Elsevier Ltd, 2020, **37**, pp. 3534–3537.

45. Gavhane, A., Patil, P., Soni, S., and Sutar, D. Utilisation of E-plastic waste in concrete. *Int J Eng Res Technol* 2016, **5**, 594–601.
46. Arora, A., and Dave, U. Utilization of E-waste and plastic bottle waste in concrete. *Int J Stud Res Technol Manag* 2013, **1**, 398–406.
47. Lakshmi, R., and Nagan, S. Studies on concrete containing E-plastic waste. *Int J Environ Sci* 2010, **1**, 270–281.
48. Mathur, A., Choudhari, A., Yadav, P.S., and Murari, K. Experimental study of concrete using combination of E-waste and plastic waste as coarse aggregate. *Int J Civil Eng* 2017, **4**, 36–39. doi:10.14445/23488352/IJCE-V4I5P109.
49. Kala, F., Nagajothi, P.G., and Felixkala, T. Compressive strength of concrete incorporated with E-fiber waste experiment findings. *Int J Emerging Technol Adv Eng* 2014, **4**. doi:10.13140/RG.2.2.13814.91207.
50. Chen, C.H., Huang, R., Wu, J.K., and Yang, C.C. Waste E-glass particles used in cementitious mixtures. *Cem Concr Res* 2006, **36**, 449–456. doi:10.1016/j.cemconres.2005.12.010.
51. Rohini, I., and Padmapriya, R. Effect of bacteria subtilis on E-waste concrete, in *Proceedings of the Materials Today: Proceedings*, Elsevier Ltd, 2020, **42**, pp. 465–474.
52. Prashant, S., and Kumar, M. Effect of partial replacement of coarse aggregates with E-waste on strength properties of concrete, in *Lecture Notes in Civil Engineering*, Springer, 2019, **25**, pp. 555–560.
53. Semiyaga, S., Acheng, P.O., Wesonga, R., Matovu, M.J., and Manga, M. E-plastic waste use as coarse-aggregate in concrete, in *Conversion of Electronic Waste into Sustainable Products. Sustainable Materials and Technology* (Jawaid, M., Khan, A., eds), Springer, Singapore. https://doi.org/10.1007/978-981-19-6541-8_6.

Index

adherend 98–100, 102
  composite 102, 109–111
  epoxy 100
adhesion 53, 139, 204, 206
adhesive layer 97, 101–102, 108–109
adhesives 9, 97–98, 100–102, 104–105, 107–111, 136, 151, 201
admixture 29, 137, 139–141, 148, 154, 162, 207
  oil-based polymer 154
  starch-based viscosity-modifying 141
  viscosity-modifying 28
aging 203, 206, 246, 250, 253–254, 256–258, 271
agriculture 2, 55, 74, 87, 216–217
air pollution 183–184
alumina 228–229, 233
aluminum 30, 85, 101, 156, 170, 262–263
aluminum oxide 219, 221–223, 225–226, 228, 231–233, 235–236
amylopectin 7–9, 135, 152
amylose 7–9, 135
application 69, 73, 75–76, 81, 85–86, 88, 95–97, 101–102, 176, 178, 188, 190, 200–202, 220–221, 227
  concreting 235
  engineering 177, 198, 200
  geotechnical 69
  industrial 136
  pavement 150
  roofing 203
  structural 134
Araldite 98, 100–101
asphalt 199, 201–203, 205–207
asphalt mixture 198, 203, 206
asphalt pavement 198, 246–247, 258

BA *see* bottom ash
Bacillus sphaericus (*B. sphaericus*) 169, 170–179
Bacillus subtilis (*B. subtilis*) 169, 170–179
bacteria 169–173, 175–179
bacterial strain 169–176, 178–179
binder 2, 10, 14–15, 52–53, 136, 151, 197–198, 201–203, 207–208, 216, 218, 220, 225–226, 245–247, 249
  asphalt 198, 203, 207, 245, 247
  biodegradable 1–2, 148–149, 154
  by-products-based 216
  cement-based 148, 153
  hybrid 10
  polymer-modified 201
  starch 1, 10, 13, 18, 20, 151, 156, 158
  ternary 155–156
  tricalcium silicate–based 128
bitumen 66, 68, 72, 197–200, 202–208, 245–250, 252, 254–258
  modified 205, 251–253, 255–257
bitumen modification 197, 203, 205, 207, 246, 257
bitumen modifier 206–207
bitumen viscosity 204, 250
blend 138, 246, 248–249, 253–258

heterogenous 208
viscous 256
blending 15, 54, 155, 204, 206, 208, 255, 257
bottom ash (BA) 87
B. sphaericus see *Bacillus sphaericus*
B. subtilis see *Bacillus subtilis*

calcium silicate hydrate (CSH) 70, 84, 116, 174
carbon quantum dot (CQD) 148, 150–151, 153–162
cashew nut shell liquid (CNSL) 207
CCA *see* corncob ash
CCNTF *see* chopped carbon nanotube fiber
cement 15, 34, 40–42, 44–45, 57, 71–72, 122–123, 137–141, 149–150, 154, 174–175, 216–219, 221–223, 225–233, 235–236, 262
composite 118
green 46–47
normal 220, 227, 232
standard 218, 220–221
cement hydration 40, 137
cement mortar 45–47, 137–138
cement paste 33, 70, 122–124, 137, 140–141, 267
cement production 149, 154, 227
CFA *see* C fuel ash
C fuel ash (CFA) 40–47
chloride 116, 122–123
chopped carbon nanotube fiber (CCNTF) 52–54, 58–60
clay 83–84, 87, 184
CLSM *see* controlled low strength material
CNSL *see* cashew nut shell liquid
coagulation 2, 248, 255–256
coarse aggregates 15, 29, 53, 170–172, 265, 267, 269
cohesive zone modeling (CZM) 96–98, 101, 103, 110
composite 1–2, 11–13, 53, 75–76, 96, 148, 153–155, 159–161, 262–263, 267, 272
banana fiber 76
bio-based 16
cement 59, 138
cement-based 133, 136, 161
compressive strengths of 159, 267
cushioning 11
epoxy binder 75
hemp-based 139
innovative 148, 153, 155–162
polymer-matrix 76, 136
pure rubber 75
rubber nano clay 76
starch-based 3–4, 139
compression 16–17, 53, 173, 177, 262
compression testing machine (CTM) 173, 177
compressive strength (CS) 16–18, 30–32, 40–41, 43–46, 136, 138–140, 148, 150, 158–159, 161–162, 173–174, 176–177, 186–193, 218, 221–222, 228–229, 231–233, 235, 267–269, 272–273
construction 74, 77–79, 87–88, 117–118, 121, 184, 188, 198–199, 206–208, 247, 256
curvilinear 82
earth-based 79
mesh 101
pavement 197, 201, 208, 257
road 198–208, 246, 250
sustainable green 133
controlled low strength material (CLSM) 69–70
corn 3, 135, 140, 150, 221
corncob ash (CCA) 34, 149, 215, 217, 221–222

coRncrete 1, 4, 10, 16–18, 136
CQD *see* carbon quantum dot
CQD2.5 148, 156–157, 159–162
cracks 11, 58, 97–98, 117, 127, 170–171, 177–179, 198, 247, 257, 269
CS *see* compressive strength
CSH *see* calcium silicate hydrate
CTM *see* compression testing machine
CZM *see* cohesive zone modeling

deformation 2, 139, 246–247
DIC *see* digital image correlation
digital image correlation (DIC) 98
durability 83, 88, 90, 117–118, 137–138, 184, 188–189, 198–199, 203, 228, 236

earthquake 40–41, 43, 47
efflorescence 186, 189–193
elasticity 75, 117, 136, 141, 207, 255–256
energy 2, 18, 40, 66, 78, 84, 86, 206–207
  net-zero 65–66, 68, 70, 72
  renewable 66
ENR *see* epoxidized natural rubber
environment 1, 51, 56, 58, 65, 69, 73, 79, 115–116, 183–184, 215–218, 221, 261–262
  aggressive 116–118, 122–123
  controlled 223
  marine 116
  temperature-controlled 220
epoxidized natural rubber (ENR) 204–205
ethylene vinyl acetate (EVA) 247
ettringite 116, 120, 122–126, 129
EVA *see* ethylene vinyl acetate
E-waste 262–273

FA *see* fly ash
fabrication process 103, 202
failure 21, 82, 97, 103, 105, 108–109, 203, 270
ferric oxide 219, 221–223, 225–226, 228–229, 231–232, 235–236
FFA *see* F fuel ash
F fuel ash (FFA) 40–47
fibers 15, 34, 52, 139, 223, 225
flash point 249, 253–254, 257
flexural strength 28, 33–35, 40–41, 43–44, 53, 139, 170–172, 175–177, 179, 269, 271–272
fly ash (FA) 28, 35, 69, 87, 89, 149, 170–171, 185, 267, 269, 273
foam 11–12, 20–21
freshwater 117–118, 125
Friedel's salt 120–126, 128–129

gel 8, 70, 84, 122–124, 172, 247
gelatinization 3, 6–7, 11, 17–18
geopolymer 15, 34, 89, 169–170
geopolymer brick 183–186, 188, 190–193
geopolymer concrete (GPC) 169–175, 177–179
GGBFS *see* ground granulated blast furnace slag
GGBS *see* ground granulated blast furnace slag
GHG emissions *see* greenhouse gas emissions
GNP *see* graphite nano powder
GONP *see* graphene oxide nano powder
GPC *see* geopolymer concrete
graphene oxide nano powder (GONP) 52–53
graphite nano powder (GNP) 52–53
gravity 118–119, 153, 157, 171, 185, 220, 245–246, 249–250, 253, 256, 258, 262, 264–265
greenhouse gas emissions (GHG emissions) 149

grinding 29–30, 35
ground granulated blast furnace slag (GGBFS, GGBS) 28, 35, 85–86, 149, 169–172, 183, 185–189, 191, 193, 216–217, 227–229
groundnut shell ash 216–217, 225–227
grouting 52–54, 56–60

high-strength concrete (HSC) 116–120, 123–125, 127–128
HSC *see* high-strength concrete
hydrated lime 55–56
hydration process 32, 58, 70, 117–119, 122, 129, 137, 141, 265
hydraulic lime 52–53, 57, 60

ignition 30–31, 55
iron slag 183, 185–187, 189, 191, 193

joint 95–99, 103–105, 107–108, 110–111
  adhesive 96–97, 99, 111
  composite 96, 99
  double-lap 95
  scarf 95
  single-adhesive 108
  tubular 96
joint analysis 100, 109
joint configuration 95–98, 101, 103
joint design 96, 108
jute fiber 2, 5, 11–15

latex 200, 203, 206, 248–249, 251–253, 256, 258
LCA *see* life cycle assessment
lime 10, 32, 54–56, 60, 71, 79, 84–86
limestone 40, 55–56, 87, 128, 229
limestone powder 216–217, 229–231
lipid 9

marble powder (MP) 40–47, 71
material 10–11, 28–29, 41–46, 70–72, 79–80, 85–90, 116–119, 121, 134, 149–151, 159–161, 171–172, 185, 198–202, 208–209, 216, 225–227, 233, 262–263, 265
  3D-printable 141
  additive 72
  adhesive 96, 133, 135
  agricultural organic 223
  agro-composite 139
  asphalt-paving 2
  binder 18, 53, 58, 74, 77, 134, 170–171, 183, 223, 231, 235
  bio-composite 16
  biodegradable 135, 201
  biodegradable bio-composite 76
  biopolymer 198
  by-product 85, 88–89
  cement-based 31–33, 40, 72, 88, 136, 141, 199
  cementitious 69, 122, 150, 215, 222–223, 228, 235
  cement-replacement 216–217, 236
  clay soil 83
  coarse 84, 272
  compressed earth 83
  concrete 117–118, 134, 150
  construction 3–4, 10, 53, 57, 65–66, 68, 70, 72, 74, 78–79, 84–85, 134, 136
  construction binder-based 34, 45
  cushioning 11, 20–21
  energy-intensive 79, 85
  fibrous 87
  fired clay 84
  involatile 203

joint 98
latent hydraulic 159
nonrenewable 198
non-rubber 201
packing 20
polymer 201, 247
pozzolanic 220–221
raw 72–74, 78–79, 136, 152, 248
renewable 198
renewable biological 136
resin-starch foam 21
sandstone 136
sensitivity clay 79
silicon-based 231
soil-based 77
soundproofing 200
starch-based 6, 14, 16
starch-based sandstone 136
steel-based 77
thermal insulation 10
unfired clay 77, 88
viscoelastic 202
waste 27–28, 35, 74, 216, 219, 223, 233, 235
mesopore 58
metakaolin 28, 128, 170
micropore 58, 162
microstructure 70, 119, 121, 124
microwave oven 16–18
millet husk ash 215, 217–219
moisture 9, 18, 52, 56, 203, 235, 247, 256
mortar 33, 40–46, 58, 71, 80, 82, 117, 121, 137, 267, 272
MP *see* marble powder

nanocarbon dot (NCD) 27–36, 40–48
nano clay 40, 74–76
natural rubber (NR) 197–209, 246–258
NCD *see* nanocarbon dot
NR *see* natural rubber

oil 66, 71–72, 201
oil sand 66–68, 71–72
oil sand waste (OSW) 66, 69, 71–72
OPC *see* ordinary Portland cement
ordinary Portland cement (OPC) 27–29, 32, 35, 42–43, 46–47, 149, 220
OSW *see* oil sand waste
oven 16–18, 204–205

palm oil fuel ash (POFA) 217, 223–225
pavement 148, 150, 153–154, 157, 162, 198–199, 201, 203, 206–209
PCB *see* printed circuit board
penetration index 250, 253–254, 256–257
penetration point 246–247, 249, 251, 254–255, 258
penetration value 247, 251, 254–255
PFA *see* pulverized fuel ash
PMB *see* polymer-modified bitumen
POFA *see* palm oil fuel ash
pollution 149, 154
polymer 6, 8, 96, 121, 133, 135, 199, 202–204, 206, 208, 247
branched 9
flame-retardant 263
helical 135
polysaccharide 134
shape memory 171
straight 8
polymerization 8, 139, 159, 161–162
polymer-modified bitumen (PMB) 199, 207–208
polypropylene resin 4, 11, 20–21
pores 58, 85, 119–120, 126–127, 174

capillary 58
intergranular 126–127, 129
isolated 126–127, 129
spherical 58
porosity 14–16, 52–53, 57–59, 84–85, 116–117, 119–122, 126, 128, 150, 265
Portland cement 1–5, 10–15, 40–46, 52, 73–74, 79, 84–85, 87, 126, 134, 136
portlandite 40, 55, 120, 122–124, 126
printed circuit board (PCB) 263
process
combustion 233
firing 83–84, 89
hardening 57–58
industrial 215–216, 232
properties
adhesive 198, 201–202
chemical 30, 55, 152, 200, 220
concrete 133
crack resistance 198
dynamic 70
elastic 247, 255
engineering 171
film-forming 151
heat-insulating 14
hydrophobic 201
microstructural 118, 120, 129
non-destructive 52
permeation 140
pozzolanic 219–220
rheological 134, 137–138, 140, 204
self-curing 151
self-healing 170, 177, 247
shear 101
thermoplastic 203
viscoelastic 141
viscosity-modifying 156, 159
water-resistant 201
pulverized fuel ash (PFA) 86–87
quarry dust 216–217, 235–236
quicklime 55–56

ratio
alkaline binder 170
binder-to-aggregate 53
binder-to-sand 57
high strength-to-weight 201
iron slag/GGBS 186–187
starch/hemp 139
straw–gypsum binder 16
water-to-binder 46, 53
water-to-cement 120
water-to-lime 58
water-to-starch 3
recycling 69, 87, 262
resin 4–5, 10–11, 19–21
resin–starch foam 4–5, 19–21
RHA *see* rice husk ash
rice husk ash (RHA) 89, 148–162, 170, 185
road 85, 171, 197–199, 203, 206, 208–209, 246, 255
asphalt 199, 245, 247
rubber 75, 200–203, 205, 208, 247–248, 251–252, 254–258

SAGD *see* steam-assisted gravity drainage
sand 3–4, 10, 15, 53–54, 57, 69, 118–119, 136, 171, 173
clayey 84
fine grouting 53
fine mortar 33
river 184
sea 117
S-BDB *see* starch-based biodegradable binder
SCC *see* self-compacting composite
SCM *see* supplementary cementitious material
SDGs *see* Sustainable Development Goals
seawater 116–129

segregation 138, 159, 162, 208
self-compacting composite (SCC) 27–35, 122
self-healing concrete 170–172
self-healing method 170–171
setting time 58, 136–138, 148, 153–156, 220, 228, 233, 236
sewage sludge ash (SSA) 216–217, 233, 235
SHCC *see* strain hardening cementitious composite
silica 30, 85, 156, 170, 225, 231, 262–263
silica content 218, 220, 222, 227, 229
silica fume 28, 35, 89, 216–217, 231–232
silicon dioxide 66, 219, 221, 223, 225–229, 231–233, 235–236
silicon dioxide content 220, 222–223, 225–227, 231–233, 235–236
single-lap joint (SLJ) 95–96, 98, 109
SLJ *see* single-lap joint
sodium silicate (SS) 183–186, 189–191
softening point temperature 245–247, 249–251, 255
soil 74, 83–85, 225
splitting tensile strength (STS) 28, 32–33, 35, 148, 160–162, 172, 174, 176–178, 272
SS *see* sodium silicate
SSA *see* sewage sludge ash
starch 2–3, 5–13, 20, 133–141, 149–151, 154, 159, 161, 201
  cassava 138–139, 150
  corn 3–4, 8, 11–12, 15, 18, 136, 140
  corn-based 17, 19
  dry 10
  gelatinized 136
  gel-forming 8
  maize 135, 139
  waste-recovered 136
  waxy 11
  wheat 4, 8, 11, 15
  wheat-based 17, 19
starch-based admixture 136–137
starch-based biodegradable binder (S-BDB) 148, 151–162
starch powder 3, 8–11, 13, 15, 17, 19
steam-assisted gravity drainage (SAGD) 67, 71–72
steel 78, 116–118
strain hardening cementitious composite (SHCC) 141
strength gain 40, 46–47, 84, 159, 216, 267
stress 74, 76, 96, 100, 105, 141, 202, 246, 263
stress distribution 97, 104–105
STS *see* splitting tensile strength
supplementary cementitious material (SCM) 28–29, 32–33, 35, 149, 216–218, 220–221, 227, 229
Sustainable Development Goals (SDGs) 28, 32, 35, 262
swelling 3, 83, 151, 204, 255–256

tailing pond 68–69
tailing powder 28, 30–35
TAST *see* thick-adherend shear test
temperature rise inhibitor (TRI) 134, 137
temperature susceptibility 140, 206, 250, 253, 256, 258
tensile strength 30, 32, 75, 100, 118, 136, 139, 161, 174, 269–270, 272
thermal conductivity 1–3, 14–16, 136, 140, 216
thick-adherend shear test (TAST) 101
tobermorite 120, 122–124

toluene 208
TOSW *see* treated OSW
treated OSW (TOSW) 69–72
TRI *see* temperature rise inhibitor

universal testing machine (UTM) 32, 122, 174
UTM *see* universal testing machine

viscosity 11, 74, 154, 206–207, 245–247, 250, 252–258
viscosity-modifying admixture (VMA) 28, 32, 141
VMA *see* viscosity-modifying admixture

waste 28, 41, 66, 68–69, 71, 86–87, 135, 184–185, 218, 223, 225
  electronic 262, 272
waste marble sludge dust (WMSD) 148–162
waste oil sand ash 66, 72
water absorption 40–45, 47, 53, 59–60, 118–119, 128, 175–176, 187–192, 262, 265–266, 272
water absorption test 175, 186, 189, 191
water content 10–11, 16–18, 20, 83, 118–119, 172
water hydrophobicity 52, 59–60
water-reducing agent 134, 140–141
water retention 54, 136, 141, 216
wheat straw ash 215, 217, 219–221
WMSD *see* waste marble sludge dust

X-ray diffraction (XRD) 66, 70, 119–121
XRD *see* X-ray diffraction

Young's modulus 100